◈ The Abyss Stares Back ◈

CARY WOLFE, SERIES EDITOR

72 *The Abyss Stares Back: Encounters with Deep-Sea Life*
Stacy Alaimo

71 *Prosthetic Immortalities: Biology, Transhumanism, and the Search for Indefinite Life*
Adam R. Rosenthal

70 *The Memory of the World: Deep Time, Animality, and Eschatology*
Ted Toadvine

69 *Hermes I: Communication*
Michel Serres

68 *Nietzsche's Posthumanism*
Edgar Landgraf

67 *Subsurface*
Karen Pinkus

66 *Making Sense in Common: A Reading of Whitehead in Times of Collapse*
Isabelle Stengers

65 *Our Grateful Dead: Stories of Those Left Behind*
Vinciane Despret

64 *Prosthesis*
David Wills

63 *Molecular Capture: The Animation of Biology*
Adam Nocek

62 *Clang*
Jacques Derrida

61 *Radioactive Ghosts*
Gabriele Schwab

60 *Gaian Systems: Lynn Margulis, Neocybernetics, and the End of the Anthropocene*
Bruce Clarke

(continued on page 252)

The Abyss Stares Back

Encounters with Deep-Sea Life

Stacy Alaimo

posthumanities 72

University of Minnesota Press
Minneapolis
London

The University of Minnesota Press gratefully acknowledges the generous assistance provided for the publication of this book by a Presidential Fellowship in Arts and Humanities from the University of Oregon.

Portions of the introduction are adapted from "Unmoor," in *Veer Ecology: A Companion for Environmental Thinking,* edited by Jeffrey Jerome Cohen and Lowell Duckert (University of Minnesota Press, 2017). Portions of chapter 1 are adapted from "Violet-Black," in *Prismatic Ecology: Ecotheory beyond Green,* edited by Jeffrey Jerome Cohen (University of Minnesota Press, 2013). Portions of chapters 1 and 2 are adapted from "Deep Sea Speculations: Science and the Animating Arts of William Beebe, Else Bostelmann, and John Wickham," *Journal of the Fantastic in the Arts* 33, no. 3 (2023): 69–91. Portions of chapters 1 and 4 are adapted from "Jellyfish Science, Jellyfish Aesthetics: Posthuman Reconfigurations of the Sensible," in *Thinking with Water,* edited by Cecilia Chen, Janine MacLeod, and Astrida Neimanis (McGill–Queen's University Press, 2013); reprinted by permission of McGill–Queen's University Press. Portions of chapter 3 are adapted from "Feminist Science Studies and Ecocriticism: Aesthetics and Entanglement in the Deep Sea," in *The Oxford Handbook of Ecocriticism,* edited by Greg Garrard (Oxford University Press, 2014), http://global.oup.com/academic; reprinted by permission of Oxford University Press.

The book's first epigraph is an excerpt from "Asters and Goldenrod," from *Braiding Sweetgrass: Indigenous Wisdom, Scientific Knowledge and the Teachings of Plants* by Robin Wall Kimmerer; copyright 2013 by Robin Wall Kimmerer; reprinted with the permission of The Permissions Company, LLC, on behalf of Milkweed Editions, milkweed.org. The second epigraph is an excerpt from "You Are Here: A Manifesto" by Eileen Joy, in *Animal, Vegetable, Mineral: Ethics and Objects,* edited by Jeffrey Jerome Cohen (Punctum Books, 2012). The Introduction contains an excerpt from "Deep Water Trawling" from *Fast* by Jorie Graham; copyright 2017 by Jorie Graham; used by permission of HarperCollins Publishers. Chapter 3 contains an excerpt from the "Foreword" by Jesse H. Ausubel to *Life in the World's Oceans: Diversity, Distribution, and Abundance,* edited by Alasdair D. McIntyre (Wiley-Blackwell, 2010).

Published by the University of Minnesota Press
111 Third Avenue South, Suite 290
Minneapolis, MN 55401-2520
http://www.upress.umn.edu

ISBN 978-0-8166-3044-8 (hc)
ISBN 978-1-5179-1873-6 (pb)

A Cataloging-in-Publication record for this book is available from the Library of Congress.

Printed in the United States of America on acid-free paper

34 33 32 31 30 29 28 27 26 25 10 9 8 7 6 5 4 3 2 1

For Kai

He told me that science was not about beauty

—Robin Wall Kimmerer, *Braiding Sweetgrass*

Initial starting conditions of spontaneous acts of combustive generosity and impossible unconditionalities. Making space, without liens, for the arrival of strangers whose trajectories are unmappable in advance.

—Eileen Joy, "You Are Here: A Manifesto"

Contents

Preface

Deep-sea life is having a moment—a long-overdue moment of accelerating scientific discovery, popular regard, and political advocacy. This sudden attention to abyssal life seems paradoxical because the deepest zones of the ocean were long thought to be devoid of life, due to the darkness, cold, and immense pressure of the water column. But the depths have also been seen as an extraordinary biome capable of harboring living fossils, or species that managed to survive unchanged through evolutionary and geologic time scales. Indeed, not only do many deep-sea species live at slower temporalities, but also some elements of their habitats, such as benthic manganese nodules, develop extremely slowly—a few millimeters per million years. Such unfathomable temporalities along with inconceivable conditions for life (dark, cold, under pressure) results in thinking of the depths—if indeed one considers them at all—as another world, intact, far from the reach of anthropogenic harms. If only that were the case. Deep-sea life faces several existential challenges, such as industrial fishing and trawling, but most dramatically the impending threat of deep-sea mining, which could, at an unthinkably precipitous pace, destroy colossal areas of the seafloor, kill a multitude of living creatures, and disrupt interrelated ecosystems, perhaps even exacerbating climate change.

Against the immense threats to abyssal life that propel this book, astonishingly beautiful images of often newly discovered deep-sea animals appear through such outlets as social media, magazines, newspapers, TED talks, films, videos, art, and coffee-table books. Despite the mind-boggling remoteness of the deep, such creatures somehow seem to be suddenly at hand, immediately present for aesthetic appreciation, regard, and speculation. I have found this to be deeply weird. While the aesthetic seems a flimsy, fragile, and inadequate mode of confronting the grave magnitude of the anthropogenic harms to ocean life, it may nonetheless be vital for inspiring a sense of connection and concern for abyssal species. While an oceanic aesthetic has long been cast as sublime, propelled by views across the vast surface of the seas, here I argue for an aesthetic more appropriate for the depths, especially in an era of extinction, specifically a creaturely aesthetic of highly mediated encounters with particular species. Analyzing the science, art, and literature of deep-sea animals from William Beebe and Else Bostelmann in the 1930s to the Census of Marine Life and other works in the beginning of the twenty-first century and extending to the current moment marked by abyssal clickbait as well as more intimate mediations, this book overflows with fabulous, strange, surreal, astonishing, spectacular, intriguing, breathtaking, and affecting deep-sea life. The aesthetic, embodied, emotional responses to deep-sea creatures, as they appear within paintings, photographs, films, videos, scientific histories and memoirs, and science fiction, disorient, captivate, and inspire. I hope readers will find the aesthetic, philosophical, and ethical currents pulsing through the science, literature, and art of abyssal species to be illuminating and perplexing, inciting a sense of encounter with enigmatic abyssal beings who seem to stare back. Even as such encounters are mediated, staged, and speculative, when the abyss stares back, the occasion calls for recognition, reckoning, and a radical expansion of environmental concern. I also hope the book will be of use to scholars, artists, activists, and others who are concerned about ocean life, from the shore to the bottom of the sea. I am grateful to everyone working in deep-sea biology and ecology, marine biology, marine conservation, marine science studies, the blue humanities, critical animal studies, environmental studies, posthumanism, Indigenous studies, Black studies, Oceanic/Pacific studies, and related fields. I am also grateful for the environmental organizations, volunteers, activists, and wildlife rehabilitation practitioners (including River Alaimo) who do such invaluable work for plants, animals, and ecosystems. Multiple, overlapping modes of intellectual, creative, and political work are necessary to inspire and mobilize concern for the abyss at hand.

Introduction

Caring about the Abyss

A *New Yorker* cartoon opens *The Silent Deep: The Discovery, Ecology, and Conservation of the Deep Sea,* by marine ecologist Tony Koslow. The caption states, "I don't know why I don't care about the bottom of the ocean, but I don't"[1] (Figure 1). The middle-aged, middle-class white women, snug in their domestic comforts, inhabit a world unfathomably different from that of the deep seas. The wry contrast between the bottom of the ocean and the arid tea party lightens the woman's confession, excusing her exhausted empathy. It also suggests that environmentalists have gone too far. How deep must people's sympathies be expected to travel? How much concern can unknown, unrecognizable life-forms elicit? When considered by a marine ecologist such as Koslow, however, the cartoon suggests the challenges facing deep-sea biology and conservation. The cartoon has struck many a nerve, it seems, as deep-sea biologist Cynthia Van Dover notes that "nearly every deep sea biologist has a dog eared copy."[2] Stronger popular interest in the depths could increase funding for ocean research and conservation. Recalling this cartoon, a "senior scientist at a major oceanographic institution," calling the speaker "Mildred," responds to her lack of interest: "What motivates me is not to make Mildred happy. What motivates me is almost the romance of exploration, to know that when you're down in a submarine and you're looking out of the window, that you're the first human ever to see that."[3] The "romance of exploration" may spark interest in deep-sea exploration, but such romance is also drenched in colonial histories. While this white, middle-class domestic scene is terribly confined, it does broach the question of what it would take to motivate popular concern for distant marine environments.

Why care about life at the bottom of the ocean? Because the start of the twenty-first century is, according to deep-sea biologist Helen Scales,

"I don't know why I don't care about the bottom of the ocean, but I don't."

Figure 1. Charles Saxon cartoon, published in the *New Yorker,* captioned "I don't know why I don't care about the bottom of the ocean, but I don't." Charles Saxon/The New Yorker Collection/The Cartoon Bank.

"without a doubt a golden era for deep sea exploration,"[4] and in the words of Edith Widder, we "are poised on the brink of massive destruction of oceanic ecosystems."[5] This moment of crisis—a time when both deep-sea exploration and the devastation of oceanic ecosystems are accelerating—demands the attention of scientific experts, policymakers, activists, artists, filmmakers, writers, scholars, governments, NGOs, consumers, and publics. Even the life dwelling in the deep oceans, seemingly safe from human incursions, is precarious, as industrialized overfishing, impending deep-sea mining, heating and acidifying waters, and pollutants (plastic, chemical, radioactive, sonic, and other) harm even benthic animals and ecosystems. The title of marine biologist Helen Scales's book telegraphs the magnificent life in the deep as well as its precarity: *The Brilliant Abyss: Exploring the Majestic Hidden Life of the Deep Ocean, and the Looming Threat That Imperils It.* While making the case that environmentalists must be optimists, given the multiple, daunting threats to the continuation of life on this planet, Widder concludes *Below the Edge of Darkness: A Memoir of Exploring Light and Life in the Deep Sea* by stating, "We're left with only one option: We're going to have to 'science the shit out of this.'"[6]

Widder, a celebrated oceanographer and marine biologist, ventures into the realm of the arts and humanities when she argues that everything depends on the public reception of scientific disclosures: "Our survival on this planet depends on fostering a greater sense of connection to the living world, and wonderment is key to forging that link. I have long believed that bioluminescence provides a means to reveal the wonder in this unseen world to a public that is alarmingly unaware and, thus, largely indifferent to what makes life possible on our planet."[7] Stressing wonder, along with imagination and curiosity, transports scientific understanding of the bioluminescent depths to a public imaginary. Widder's aesthetic musings in *Below the Edge of Darkness* continue a tradition in which writers, artists, and scientists accentuate the aesthetic dimensions of the seas. As the blue humanities—the oceanic and more generally aquatic counterpart to green or terrestrial environmental humanities—develops, the question of what the aesthetic means or does as it circulates through science, culture, and art will be essential. Marine biology is saturated with stylized aesthetics that suggest the currents that run between science, sensation, and public reception. *The Abyss Stares Back* investigates how aesthetic images of deep-sea life circulate through science, scientific memoir, science fiction, art, popular culture, and environmental advocacy, mainly but not exclusively in the United States. After this theoretical introduction, the book examines William Beebe's legendary dives in his bathysphere in the 1930s, continues with the strategies of containment within science writing and science fiction of the mid- to late twentieth century, arrives in the early twenty-first century to plunge into the ambitious and aesthetically rich Census of Marine Life, and concludes in the present moment of clickbait, racist voids, intermeshed models of kinship, and mediated intimacies. The fact that visual representations of nature—be they the classic landscape paintings that shaped the constructed viewpoints of the Grand Canyon or contemporary Sierra Club calendars—have been essential for the development and popularization of U.S. environmentalism has been well established within environmental studies. The deep seas, however—which are not accessible to tourists and ramblers; which can only be experienced through highly mediated, expensive, scientific expeditions; and which are frequently labeled as alien worlds—require their own examination in terms of the aesthetic reception of the beautiful, the surreal, the strange, and the unknowable.

As a cultural and theoretical foray focusing on the travels of aestheticized creatures, rather than a marine biology text or scientific history, what constitutes the deep will vary throughout the chapters, from Beebe's "half mile down" (800 meters) in the 1930s to the current designations of the bathypelagic (1,000–4,000 meters), abyssalpelagic (4,000–6,000 meters), and hadalpelagic zones (6,000–1,100 meters), along with their benthic (bottom) counterparts. As a nonscientist discussing nearly a century of science, art, film, and literature, I refrain from imposing definitions or distinguishing between bathypelagic, abyssalpelagic, and so on. Instead, I follow my sources and echo their terminology, often alternating between "deep sea" and "abyss." The term "abyss" is certainly not neutral, as it resonates with the cultural imaginaries of the deep. While the depths are widely understood to be exceptional habitats with extraordinary creatures, the most capacious definition of the deep ocean asserts that it begins at two hundred meters down, where darkness begins, which means that it includes 90 to 95 percent of the volume of the ocean, making it the largest biome on the planet. Alan Jamieson and coauthors argue for a less sprawling definition, restricting the deep sea to the waters below a thousand meters, drawing this line so as not to "confuse the matter with habitats, species, and anthropogenic impacts that though they might be present in the deep sea, are typically found in the upper layer."[8] In any case, the deep seas are neither eccentric nor negligible. The Deep Sea Conservation Coalition definitively declares that "all life on earth, including human life, depends on the deep sea."[9] As life itself hangs in the balance, the deep sea epitomizes this moment, termed the Anthropocene, when the scale of anthropogenic harm overwhelms the political will and the ability to change course.

While this book ponders the magnitude of the deep-sea biome, especially in terms of scientific captures, speculative mapping, posthumanist unmooring, and Anthropocenic scale shifting, deep-sea animals, in their shimmering multiplicity, steal the show, graciously and metonymically extending their magnetism throughout their biome.

The Abyss Stares Back brackets marine mammals and other so-called charismatic megafauna, focusing instead on animals such as hatchetfish, siphonophores, and cephalopods. Many of these animals not only have little history within human cultures but are also profoundly different from what humans (as terrestrial mammals) expect animals to be. Jellyfish and other gelatinous animals whose bodies are 95 percent water, for example, float at the far reaches of our ability to construct sturdy interspecies connections, posing both conceptual and ethical or biopolitical challenges. What sort of ethical response to gelatinous creatures is possible when even some of the most esteemed and beloved mammals, such as cetaceans and our fellow primates, struggle for survival? Because most of the ocean is inaccessible to humans and abyssal life is extraordinarily diverse, deep-sea creatures seem beyond the reach of human comprehension and responsibility. Are deep-sea animals simply too distant and too strange to spark concern? Perhaps. Yet aesthetic encounters, even those that are highly mediated, can dodge conventional alienation from abyssal life. The aesthetics of the breathtakingly beautiful, adorable, diaphanous, radiant, weird, and surreal pulse with a kind of intimacy, sparking an emotional sense of connection that in turn ignites curious speculations about species' being and creaturely lifeworlds. Abyssal life may seem to exist worlds away, yet the experience of an aesthetic encounter seems immediate, affecting, and potent. The abyss stares back in depictions, framings, stagings, and designs that are anything but disinterested.

Despite the abundance of dazzling images of deep-sea life traveling through marine science, art, and popular culture in the twenty-first century, which call publics to imagine abyssal worlds, both cultural and scientific conceptions of the deep seas are rather recent. As Nicole Starosielski notes, "the ocean did not always have depth or volume in the popular imaginary." By attending to the history of undersea cables, she argues that although "all forms of depth are mediated in a general sense by cultural forces and specifically by instruments and representational technologies," the "depth of the ocean was critically mediated by network infrastructure in both of those senses."[10] Mediations of the depths involve infrastructure, technology, and the practices of science. The deep sea has long posed formidable problems for science, as it has been difficult to capture "specimens" and, until relatively recently, impossible to observe species in their habitats and ecosystems. As Renisa Mawani explains, the very "materiality of oceans—their changing temperatures, moving currents, and dynamic forces—posed a significant challenge to technological innovation and human mastery."[11] This is even more the case for the oceans' deepest zones. Without much information about species and their interrelations, the initial recognition of abyssal life is often highly aesthetic. However, the aesthetic response—which can be emotional, personal, and unsettling—can complicate conventional scientific epistemologies. Paradoxically, the beauty of deep-sea life has both threatened scientific authority and amplified scientific reception. When at the start of the twentieth century William Beebe resisted the strict separation of art from modern science, he was critiqued for being a mere showman, not a proper scientist. However, by the turn of the twenty-first century, the Census of Marine Life featured and finessed the aesthetic dimensions of their findings, and nearly a quarter of the way into the twenty-first century, a creature feature such as *Meg 2: The Trench,*[12] poses scientists against capitalist extractivism, aligning them with an aesthetically potent (albeit campy) conservationist politics. Even as aesthetic encounters disrupt standard scientific epistemologies, they can nonetheless provoke scientifically informed speculations and foster public imaginaries necessary for marine conservation.[13] What seems to be an epistemological or scientific failure can instead be a recognition of the magnitude of biodiversity that swamps human knowledge systems and inspires more capacious, philosophical, aesthetic, and perhaps ethical relations with abyssal lifeworlds. Such epistemological failures may check the Western drive to master, objectify, contain, and flatten an externalized nature, instead prompting a volumetric vision of the astonishingly heterogenous forms of life on this planet. This book traces the cultural work of creaturely aesthetics as they flow through science, art, literature, film, and popular culture, asking how they disrupt conventional scientific epistemologies, how they are implicated in a colonizing environmental gaze, how they populate abyssal voids with beings that spark attachment and concern, and how they function within the Anthropocenic horizons of extinction. What do we see when we see a siphonophore? What can such an image do? What would it take for such an image to inspire more capacious and potent ecological visions?

Casting Around in the Void: Abyssal Knowing and Being

The rather sudden appearance of deep-sea life, circulating through terrestrial human media, is remarkable. While medieval maps warn "here be monsters" on the open seas, featuring lively images of dragons and other fantastical creatures attacking ships or lounging on the water's surface, who knew what lurked far below? In the mid-nineteenth century, the azoic theory of British naturalist Edward Forbes declared that *nothing* could lurk below, as the extreme cold, darkness, and pressure of the deep seas prohibited life. Helen Rozwadowski notes that by the 1860s, "hydrographers, deep-sea fishermen, and whalers had found evidence of organisms at great depths," yet naturalists of the time seemed oblivious to this evidence, as they lived in different "intellectual and social worlds" from the seamen.[14] When the repair of Mediterranean cables in 1860 brought sea animals to the surface from about a thousand fathoms, "a multitude of unfamiliar creatures" was discovered, and knowledge of them was disseminated to "men of science" as well as publics.[15] Nonetheless, the technological, logistical, and financial barriers to scientific exploration of the depths have been formidable, especially in the case of gelatinous creatures, which, when dredged up as specimens, become unidentifiable mush. Yet even after evidence of life in the depths surfaced, the conception of the abyss as a void has endured with the potency of myth. Moreover, the volumetric expanses of the ocean have been flattened by capitalist enterprises; they are imagined as a merely horizontal plane for transportation, as Philip E. Steinberg argues in *The Social Construction of the Ocean.*[16] While capitalism has flattened the conception of the ocean, rendering the depths immaterial, the deep seas, conceptualized as an abyss, have long been drenched in racist formulations of Blackness. Calvin L. Warren, in *Ontological Terror: Blackness, Nihilism, and Emancipation,* argues, discussing the philosophy of Martin Heidegger, that "nothing is the essence of science—the void, the abyss, the unruly thing is the repressed ground of scientific inquiry." He continues: "How do you quantify nothing? How do you render nothing tangible, an object for observation."[17] Provocatively, he argues that "Blackness enables a scientific encounter with the horrors of an entity that is nothing and something at the same time."[18] Chapter 4 will draw on Warren's arguments to make sense of the nonsensical episode of David Attenborough’s famous 2017 documentary, *Blue Planet II,* which depicts the depths as horrifically black yet somehow simultaneously “nothing and something at the same time.” Indeed, many of the clickbait depictions of “weird” or “monstrous” abyssal life could be read as dog-whistling disavowals of, as Warren would put it, “black ~~being~~.” By contrast, the work of Nnedi Okorafor and Alexis Pauline Gumbs, also discussed in chapter 4, depicts ocean life as allies, kindred, or intermeshed being, never alien and never abject.

Casting the abyss as an emptiness where knowledge meets its end and being becomes unfathomable elicits epistemological drama. The ocean surfaces in unlikely places, enlisted as a metaphor for an unknowable void. W. J. T. Mitchell, in *What Do Pictures Want? The Lives and Loves of Images,* for example, writes: “We theorize to fill a void in thought, we speculate because we don’t have an explanation or a narrative; and so we cast a hypothetical net into the sea and see what swims in.”[19] We are plunged, metaphorically at least, into the abyss, where theory begins as a “void in thought,” an opening into what we don’t understand, can’t conceptualize, and fail to verbalize. Such failures, often performed, as Margaret Cohen argues, as the “underwater *je ne sais quoi,*” are striking in themselves, especially during moments when scientists, such as William Beebe, whom we will encounter in chapter 1, are undone by aesthetic awe.[20] Mitchell notes that his titular question—what do pictures want?—“has overtones of animism, vitalism, and anthropomorphism,” cautioning that the “epithet for our times” is not “ ‘things fall apart,’ but an even more ominous slogan: ‘things come alive.’ ”[21] While things coming alive may be the stuff of horror films, it is also a posthumanist, new materialist corrective to the deadening of the world through capitalism, colonialism, Enlightenment dualisms, the dwindling number of plants, animals, and other living beings, and the proliferation of concrete, asphalt, and plastic.[22] More playfully, we can read the animacy in this passage in terms of an ontoepistemology in which material and multispecies agencies propel knowledge or provoke theory. If you cast your net of not-knowing, something fabulous might just swim in.

Such frameworks evoke epistemological humility and muddle the conceptual chasm between natural reality and cultural representation as well as between land and sea. Given the scale of the global seas and the limits of scientific knowledge about marine creatures and ecologies, the blue humanities and marine science studies could be considered as a fluid process of catch and release: something swims in, but it must be released back into the intraacting flows, relations, and systems that demand Anthropocenic scale shifting along with an assessment of harms, entanglements, and modes of fierce, politicized care. "Capture," with its undertones of cruelty, objectification, and animal resistance, is a candid term for thinking through both the scientific and aesthetic disclosures of deep-sea creatures, as it conveys both an ontoepistemology of the process of accessing and understanding a living world as well as the ethicopolitical responsibilities inherent within these processes. "Capture" is a key term throughout this book because of its multimodal valences. It can be understood literally in terms of the violent seizure of live animals as specimens; more methodologically or theoretically as an aspect of "mangled" scientific practice that discloses a material reality, to use Andrew Pickering's term; and as photographic or video capture, in which a moment in place and time is rendered into an image—a "circulating reference," in Bruno Latour's terms, that retains something of what was captured.[23]

"Capture" might lead us off course, however, as it could imply mastery, objectification, and epistemological stasis, or the separation of subject from object. Captures could be understood in terms of more fluid notions of what it is to see and to know within networks of human knowledge practices and technologies as they interact with nonhuman agencies. Against the buoyant quote by Mitchell, we could consider D. Graham Burnett's more agonized attempts to think with cetaceans in ways that are not merely metaphorical. Burnett concludes his massive, extensively researched, nearly eight-hundred-page volume *The Sounding of the Whale: Science and Cetaceans in the Twentieth Century* by stating, "The fundamental lesson I have taken from the research and writing of this book amounts to nothing less than a kind of sweeping epistemological humiliation."[24] What is "exportable" about his undertaking is, he contends, an "anti-analysis" he has "not figured out."[25] While a recognition of epistemological limits has been articulated as an ethical stance within feminist, environmental, and decolonial theory as a means of countering colonizing epistemologies of mastery, Burnett seems stranded within his own strict division between archives of whale sciences and actual whales. He warns us, early on, that there will be no whales in this book but "only words about whales."[26] He continues, "What were the whales saying? I have no idea. Do I give too much agency to (human) words? Maybe. It is ever thus with bookish folk. If it is whales you want, you have to go to sea."[27] As tempting as it is to set out to sea to encounter whales, not words, the larger epistemological problematics posed by what it means to apprehend most oceanic species cannot be so readily resolved, as all encounters, even those at sea, are always already mediated and would not yield the solid knowledge that Burnett seems to seek. The multitude of historical, mythic, scientific, literary, and philosophical musings on whales in Herman Melville's novel *Moby-Dick,* for example, unmoor readers from the drive of the plot, which includes direct and gory contact with whales, dispersing knowledge into never-ending provisional disclosures, speculations, and creations. On page 675, Burnett admits, "Knowing things is hard." True, but perhaps the critics of Western epistemologies and scientific objectivity would add that "things"—solid, factual objects—should not be the target. The inability to pin down, say, what whales are is a failure worth emulating because it dramatizes the limits of epistemologies that distance, circumscribe, and objectify. I should add here that even though I use the term "species" throughout this study, "species" must be read as *sous rature* because evolution entails both the interrelation of species and their transformation. Taxonomic captures and cuts cannot be but provisional and arbitrary, given not only the common ancestry of living beings but also the many different definitions of "species" within and beyond Western science.

While Burnett mocks the idea of giving whales "agency," sequestering the term with scare quotes, a more generous engagement with feminist science studies, starting perhaps with Donna J. Haraway's prodigious scholarship,[28] could foster something besides "a sweeping epistemological humiliation"—or, even better, could alchemize humiliation into an environmental ethics of humility. For example, Mette Bryld and Nina Lykke, in their 1999 volume *Cosmodolphins,* begin by aligning themselves with feminist cultural studies scholars who share "a disbelief in the traditional dichotomies between theory and empirical objects of study, between knowing subjects and objects of knowledge."[29] Similarly, Donna J. Haraway, in a chapter on "Tentacular Thinking" in *Staying with the Trouble: Making Kin in the Chthulucene,* asserts: "The tentacular ones tangle me in SF. Their main appendages make string figures; they entwine me in the poesis—the making—of speculative fabulation, science fiction, science fact, speculative feminism, *soin de ficelle,* so far."[30] By listing multiple meanings of sf, Haraway suggests their distinctiveness and relations. There is no gulf here between words and whales but rather tentacular tangles, inspiring academic work that is speculative, sometimes even fictional. There is a rich generativity in this mess. Moreover, Burnett's epistemological or methodological white flag—his witty confessions of failure—can be read more generatively as an aestheticized outburst of awe, not unlike many of the reactions to deep-sea creatures that will be discussed throughout this book, most notably those of William Beebe.

Feminist epistemologies, at least since Carolyn Merchant's *The Death of Nature,* have critiqued Enlightenment models of knowledge for their gendered paradigm of "penetrating" a feminized nature.[31] Luce Irigaray, in *Marine Lover of Friedrich Nietzsche,* inhabits such gendered dualisms to subvert them, granting the ocean sovereign vastness. She notes that "the loftiest gaze does not penetrate thus far into her depths" because she is "much deeper than the day ever conceived her to be."[32] The trope of the ocean as unfathomably vast, however, can place it beyond the scope of environmental concern. Even Rachel L. Carson believed—at least until 1951, when *The Sea Around Us* was published—that the ocean was too immense for anthropogenic harm. Moreover, the conception of the sea as unknowable because of its magnitude can be detrimental. Irus Braverman and Elizabeth R. Johnson explain in their introduction to *Blue Legalities: The Life and Laws of the Sea* that the oceans have historically been "characterized by inaccessibility and indeterminacy." Still today, "ignorance remains central to the seas' legalities": "In the legal literature, the opacity of the oceans is most often understood to incapacitate managers of marine resources or conservationists who seek to curb pollution and battle other perils."[33] They underscore an essential point: "One of the challenges for blue legalities is figuring out how to insist on accountability and justice in the absence of complete knowledge."[34] The precise articulation of what it means for the abyss to be unknowable matters, as magical thinking poses the depths as a separate realm where human harms dissolve into invisibility. Astrida Neimanis warns, writing about chemical waste in Sweden's Gotland Sea, that conceptualizing the ocean as "universal solvent" and "pure alterity" enables us to ignore responsibility for anthropogenic harms: "Those matters swallowed up by the sea become part of its unknowable abyss—not only forgotten but rendered unintelligible."[35] As we encounter different aesthetic and scientific captures of deep-sea creatures, the question of what it means for the depths to be unknowable will repeatedly arise—as a way to dodge legal and financial responsibility, as an admission of scientific or scholarly failure, as a pervasive cultural trope, as a mathematical impasse, as an impetus for environmentally ethical epistemologies, or as an ordinary, even clichéd, sense of the wondrous or sublime.

The Sublime: Surface Void or Posthuman Provocation?

If there is one aesthetic category persistently associated with the ocean in Western art and philosophy, it would be the sublime. Yet *The Abyss Stares Back* veers away from the sublime of Immanuel Kant, Edmund Burke, Caspar David Friedrich, or even J. M. W. Turner. As a cultural studies scholar, I intend to remain open to what swims in. In other words, I attempt to read texts on their own terms rather than confining them within already established philosophical conceptions. As a feminist scholar, I hope that openness to the sources at hand offers possibilities for dodging rather than repeating dominant concepts within the Western canon of philosophy, literature, and art. Furthermore, at this point, multiple definitions and renditions of the sublime have made the term contradictory and diffuse. Although the concept of the sublime is sometimes applied to contemporary art, it seems more relevant to the eighteenth and nineteenth centuries. This project, however, begins in the 1930s and continues to the present moment—a moment in which technologies of aesthetic and scientific capture and circulation differ from those of past centuries. Finally, Kant's emphasis on aesthetic judgment as disinterested does not provide a hospitable habitat for contemporary concerns about extinction and biodiversity. Disinterest is inimical to tracing how aesthetic encounters can spark creaturely attachment and deeper environmental concern.[36]

In terms of the blue humanities, it may be useful to note, however, that not only landscapes but also seascapes have been depicted as quintessentially sublime. Sublime seascapes like Turner's, however, usually represent the stormy *surface* of the ocean, with ships being tossed about by tremendous winds and waves, dramatizing the powerful forces of nature rather than depicting marine life in sympathetic modes. In effect, the surface of the sea is cast as a kind of elementally potent void; air, water, waves, and wind toss turbulently, but marine life is scant. Moreover, some meanings of the word "sublime" manifest vertical hierarchies that privilege the sky over the depths, such as this example from 1633: "As clouds . . . being elevated and sublimed toward the upper regions of the aire, are rarefied"—or this example from 1845: "Thoughts rise from our soul as from the sea the Clouds sublimed in Heaven."[37] Abyssal aesthetics must descend rather than ascend, saturating marine life with inherent value and scrambling the vertical semiotics of good versus evil, heavenly versus hadal. Even though the vastness of the depths is consonant with prevalent conceptions of the sublime, pelagic and benthic realms do not lend themselves to seascape depictions, given their darkness, their staggeringly volumetric immensity, and, in the case of pelagic or open seas, the lack of geological features. That word, "staggering," paired with "immensity," catches me up in the inescapable stickiness of the sublime. True confession: I do evoke that sort of sublime throughout the book, but with an orientation toward posthumanism rather than humanism or Romanticism. The recognition of abyssal realms as so astonishing as to destabilize reason addresses how the aesthetic potency of deep-sea life exceeds the limits of thought and the conventional parameters of scientific epistemologies. While the sheer magnitude of the abyss is important to reckon with, however, much of this book emphasizes a creaturely aesthetics that circulates in heterogenous forms. Abyssal aesthetics includes the beautiful, the adorable, the surreal, the weird, the monstrous, the grotesque, the psychedelic, the unfathomable, and even the self-reflexively Anthropocenic. Divergent creaturely aesthetics play out within larger frameworks pertaining to scientific practice, taxonomic framing, epistemology, mediation, containment, objectification, intimacy, care, concern, kinship, responsibility, pleasure, captivation, awe, and disanthropocentrism. One aesthetic term—even a supremely protean one—cannot encompass scientific and aesthetic encounters with abyssal life.

Global Visions

Extending global environmental concern across immense pelagic realms and down to the seafloor requires volumetric visions. The coffee-table cartoon, worlds apart from such visions, expresses a stubborn commitment to disinterest. What would it mean, what would it take, for people to become interested in life on the bottom of the ocean? Would the interest of these ostensibly unmarked people be permeated by what Aileen Moreton-Robinson terms the "white possessive"? How do the histories and practices of scientific exploration,[38] settler colonialism, commodifying taxonomies, and genetic bioprospecting saturate depictions of the otherworldly creatures of the deep? We might pause to consider the iconic Captain Nemo, the man without a country in Jules Verne's *Twenty Thousand Leagues under the Sea*,[39] as he takes whatever creatures he wants for food, energy, or aesthetic pleasure, epitomizing the fantasy of unencumbered voyages that grant unmediated and unlimited possession of marine life. While Verne's taxonomic descriptions of Nemo's collections appear as lackluster lists of species, in the early twenty-first century, stunning images of recently discovered deep-sea creatures circulate, glowing on our computer screens and shimmering in coffee-table books. Highly mediated, contemporary digital images of wondrous deep-sea creatures echo the *wunderkammer*, or cabinet of curiosities. Captured in photographs, the creatures captivate their viewers, who are rapt with wonder and curiosity, affective and cognitive states that open us to fresh contemplations of the world. As beings extracted from their aquatic zones, framed as aesthetic specimens, they raise questions about Anthropocene visions, mapping, and mediation even as the aesthetic encounter shimmers with an impossible immediacy, provoking speculation about the animals' habitats in terms of scale, depth, volumetric expanse, and water column pressure.

The Anthropocene would seem to demand that we learn to scale up as climate change, the sixth mass extinction, pollution, ocean acidification, and other environmental crises are wickedly global problems. Naomi Oreskes, in "Scaling Up Our Vision," a beautifully lucid history of marine science and climate change, proclaims, "Our future will depend not only on understanding our relationship to the world ocean . . . but also on finding some means to change that relationship. It is time for us to take on the scale of the ocean—the scale of the planet—in our thinking and, in doing so, scale up our imagination of the human. It is time to scale up our vision."[40] What forms such a scaling up would take, whose perspective it would install, and what life it would make visible are questions that render scaling up anything but simple. In the penultimate chapter of my book *Exposed: Environmental Politics and Pleasures in Posthuman Times,* "Your Shell on Acid: Material Immersion, Anthropocene Dissolves," I critique this sort of scaling up, arguing that the predominant visual images of the Anthropocene attempt to represent the enormity of temporal and geographic scale by zooming up and away from the planet, epitomizing what Donna J. Haraway called the "God's eye trick" of ostensibly objective Western epistemology.[41] In this predominant mode of visualizing the Anthropocene, as well as in Dipesh Chakrabarty's influential essay, "The Climate of History," the human knower becomes an abstract, transcendent, disembodied creature who surveys the world he has affected from a safe distance.[42] This human who sees and knows stands apart from the material flows of the Anthropocene and the systems of privilege and precarity. This erases differential culpabilities and vulnerabilities of particular human groups. The formulation of the transhistorical human, the *anthropos,* acting on the planet as a geological force, often ignores biological and chemical alterations of the biophysical, ecological world. Moreover, nonhuman species vanish from sight, as they are almost never depicted in iconic Anthropocene visualizations. It is as if the sixth mass extinction had already concluded, leaving no species other than the human.

Macarena Gómez-Barris, in *The Extractive Zone: Social Ecologies and Decolonial Perspectives,* describes similar vertical visions as an "extractive view": "colonial visual regimes normalized an extractive planetary view that continues to facilitate capitalist expansion, especially upon resource-rich Indigenous territories." In short, "vertical seeing normalized violent removal."[43] As Haraway argues, the view from nowhere distances the knowing subject from the object of knowledge, objectifying a passive "nature," as a "resource."[44] Gómez-Barris, analyzing a film by Carolyn Caycedo, which "draws from Indigenous relational understandings of land," proposes a "fish eye perspective," an example of how "Global South epistemologies and philosophies of race and racism . . . differently imagine knowledge and perception as the foundation of planetary inhabitance."[45] David A. Chang (Native Hawaiian), in *The World and All the Things Upon It: Native Hawaiian Geographies of Exploration,* explains that in "Hawaiian grammar and Hawaiian discursive practice . . . one almost always speaks from a place," and "this usage structurally preserves the perspectivalism at the heart of Kanaka geographic thought."[46] The view from above not only makes it seem as if the unmarked knower floated in the sky but also distances the knower from the known, as if an abyssal emptiness stretches between the two. By contrast, Karin Amimoto Ingersoll's *Waves of Knowing: A Seascape Epistemology* emphasizes immersion, interaction, and transformation, epitomized by the practice of surfing. Her Kanaka Maoli (Native Hawaiian) seascape ontoepistemology "evolves as an interactive and embodied ontology; a kinesthetic engagement and reading of both the physical and metaphysical simultaneously, enabling an alternative epistemology for Kanaka."[47] In an analysis of the work of New Zealand Maori author Keri Hulme, Elizabeth M. DeLoughrey writes that "the use of Indigenous ontologies in relation to more-than-human nature, particularly the creatures of the sea, offers a vital critique of neoliberal extractivist regimes that are undermining Maori sovereignty of the foreshore and seabed."[48] All of these arguments offer potent critiques and alternatives to the extractive view, yet it would be appropriative for settler colonialists to take up (the common academic parlance betraying hierarchical verticality, use, appropriation, and theft) Indigenous philosophies and traditional ecological knowledges, especially given how deep-sea creatures are made present through the mediations of big science.[49]

We have drifted far from the coffee-table scene, yet Gómez-Barris, Chang, Amimoto, and DeLoughrey underscore alternatives to a universalized, transparent perspective on "the world"—a world that awaits attention. The moment one would choose to care or not care while sitting still betrays an expansive, commodifying, and often visually constructed sense of entitlement. Moreton-Robinson argues that an unmarked global vision is undergirded by "a white possessive logic." If, as Moreton-Robinson writes, "white subjects are disciplined, though to different degrees, to invest in the nation as a white possession that imbues them with a sense of belonging and ownership,"[50] then is the expansion of environmental concern, extending to the very depths of the sea, itself an extension of white possessive logic? As a middle-class white woman who grew up in Michigan on Ojibwe or Saginaw Chippewa lands, I am implicated in global visions that deliver marine creatures as aestheticized life-forms that elicit concern. Enthralled by Jacques Cousteau TV specials, learning to bodysurf during childhood visits to Florida, then snorkeling and scuba diving as an adult, my love of ocean life has been mediated by television, tourism, and (ableist) adventure culture, as well as by environmental ethics, love of marine life, and concern for all living creatures.

Kinship with marine life manifests ontological interrelation, a more intimate interconnection across species lines compared to that of conventional modes of environmentalism that externalize "the environment." Inuit stories of Sedna highlight kinship as the foundation of care for marine life, a relational mode that will also be discussed in chapter 4.[51] Sedna, whose father, to save himself, cut off her fingers as she clung to his boat in a storm, illustrates a particularly striking counterpoint to the coffee-table scene. Her fingers became marine mammals, and, in the words of Jace Weaver, drawing on Laura Adams Weaver, in turn drawing on accounts of Inuit peoples as recorded by Franz Boas, "Sedna was deified," becoming "the mistress of the Sea, responsible for her children, the marine mammals who sprang from her body. If the Inuit anger her, she will withhold the sacrifice of her children, and the Inuit will starve. A prime responsibility of Inuit shamans is to travel to the bottom of the ocean and comb Sedna's hair—because she has no hands—and keep her from becoming displeased."[52]

Even in this brief retelling, a rich sense of ethical relations, intimate kinship, and ecological responsibility resonates. Artist Ningiukulu Teevee (Kinngait) has created several depictions of Sedna, including "Sedna's Creation," in which a hand with cut fingers, placed in the middle of the work, radiates into a fluid array of lively marine mammals. In her lithograph "Sedna's Wonder," a curving half-fish, half-woman figure, underwater, reaches up to touch and marvel at a jellyfish, the waters gorgeously blue.[53] By contrast, Teevee's black-and-white "Untitled (Sedna by the Sea)" depicts a despairing Sedna, sitting alone on a rock, without her marine mammal kin, smoking a cigarette, watching garbage being dumped into the sea.[54] Alison Cooley calls the drawing "a grim view of a world where industry dominates and the seas suffer so deeply that their goddess is forced to abandon them."[55] The composition suggests an inversion of extractivism's verticality, as the garbage trucks at the top of the page dump waste that pours into the waters in the middle of the page, sinking down to where Sedna sits—on land, but also, more figuratively, at the bottom of the page and the sea. The bleakness of this black-and-white drawing contrasts with many of Ningiukulu Teevee's other works, which are brightly colored, featuring animals, people, and hybrid figures swimming or otherwise being in or near the water.

The possessive logic that suffuses vertical, extractive visions of the world whitewashes the banal processes of capitalist consumption and waste, where harms are rendered invisible and untraceable, except through explicitly activist mappings of, say, how plastic bags kill ocean life, industrial fishing harms Indigenous peoples, and electronic waste harms peoples in the global south. These long transcorporeal maps of culpability begin in human terrains, tracing the strange scales in which quotidian practices result in colossal violences. Even those who refuse to consume ocean life caught by horribly wasteful industrialized fishing, reduce the use of plastic, and shun cruise ships still purchase things that contain resources extracted through industrialized fishing and mining. Much garbage, even when supposedly recycled, ends up in the belly of a fish, seabird, or whale—not to mention the deadly effects of sonic pollution on marine mammals; the looming catastrophes caused by industrialized people's release of carbon dioxide, such as acidification, melting polar ice caps, and hotter oceans; marine extinctions; and the degradation of ocean ecologies. As ocean life is assaulted by global capitalism, colonialism, climate change, and pollution, the effects on certain groups of people, such as fishing communities and Pacific Islanders, have already been horrifically uneven.

In *African Ecomedia,* Cajetan Ikheka analyzes a potent series of photographs by Fabrice Monteiro, who lives and works in Dakar, Senegal. This series of photographs, *The Prophecy,* features majestic female figures, including versions of Mami Wata, who wear gorgeous dresses made from garbage. Iheka argues that by "staging bodies of waste, remnants of animal bodies, and other materials exhumed from the sea, Monteiro reestablishes a crime scene, one that indicts humanity for unbridled consumption."[56] Such environmental art provokes the viewer into tracing long lines of responsibility, both temporal and geographic, within capitalism and colonialism. Depicting culturally immersed scenes from different continents, Teevee and Monteiro reveal entangled harms to specific groups of people as well as to ocean life, underscoring the power of art to keep viewers from just looking away. Michel Serres in *Malfeasance: Appropriation through Pollution,* proposes an aesthetic of "dis-appropriation": "What if seeing the world's beauty—and that of human works and bodies—would merely consist in removing the waste of appropriation? To discover: to take away this covering, this deluge of garbage."[57] Such an aesthetic of removal would ironically conceal and improperly dispose of the crime scene that Iheka dramatizes. Monteiro's photographed figures, splendidly arrayed in oceanic garbage, including a thick coating of what looks like black oil, provoke a more vexing aesthetic response, their striking beauty interlaced with a disturbing recognition of harms, networks, and complicities. It is worth noting here, however, that the contemporary photos of stunning deep-sea creatures, which will be discussed in chapter 3, circulate as "dis-covered" in Serres's terms, scientifically and aesthetically captured and framed, but presented as belonging to themselves, not the viewer.

The shift Serres proposes, from Kant's disinterested aesthetic to a "dis-appropriated" aesthetic, would seem to directly counter the long history of the "white possessive," as it intends "the dispossession of the world."[58] Such an aesthetic, however, may operate as a settler "move to innocence," a "dispossession" that seems merely philosophical or metaphorical,[59] whereas a recognition of entanglement within global networks of harm can counter dominant narratives of innocent deep-sea discovery and adventure. Take, for example, the highly publicized expeditions into ocean depths by extremely wealthy white men—James Cameron, Victor Vescovo, Ray Dalio, as well as the five men who died in the *Titan* submersible implosion in 2023.[60] In the wake of the *Titan* implosion, science journalist William J. Broad defended such touristic descents, contending that the "adventure factor" helps "generate wide appreciation among the public for the wonders of the world's oceans."[61] Given that the *Titan* descended for the thrill of touring the wreckage of the *Titanic,* I doubt the trip sparked interest in the marvels of marine life or cultivated concern about anthropogenic assaults on marine ecologies. Moreover, the prevalent and weirdly anachronistic discourse of heroic deep-sea discovery revels in the "innocence" afforded by realms that are uninhabited by humans, constructing a convenient *aqua nullius* seemingly untouched by unappealing colonial histories of explorers and conquerors. The expeditions of Cameron and Vescovo can be understood in terms of Tiffany Lethabo King's argument that the "production of the White conquistador-settler is an ongoing process of violent autopoesis that must be continually rewritten and revised."[62]

The racist structure she analyzes on an eighteen-century map is not unlike the dominant vertical and disembodied vision that I have been critiquing in which "the privileged position of humanity is that which remains beyond the realm of embodied visuality."[63] In this instance, however, I would stress that what needs to be seen isn't the corporeality of these twenty-first-century explorers but rather the way their wealth affords them an unremarkable possessive relation to everything on the planet—a planet that they harm disproportionately by "virtue" of their extreme wealth. They need not settle anywhere; they need not directly harm anyone to epitomize their lineage as conquistador-settlers if we trace the trails of slow violence[64] radiating from the ways in which they exploit the systems, materials, ecosystems, and lives at their disposal.

Biological and ecological sciences of the depths also reiterate problematic global visions. The counting of species, assessment of biodiversity, and concern for marine ecologies cannot be separated from histories and ongoing practices of colonial taxonomy, biopiracy,[65] and genomic bioprospecting. While the scientific and popular imaginaries of deep-sea creatures do not reckon with colonial histories, they do seem to counter capitalist and extractivist relations to the oceans by populating an imagined ocean with unimaginable creatures, as two of the Census of Marine Life's coffee-table books suggest. The cover of *World Ocean Census: A Global Survey of Marine Life* features a bright orange jellyfish, glowing against a dark violet background, positioned in a way that suggests a faceless head facing the viewer, daring viewers to see this creature as a living being worthy of concern. Even a creature without eyes can seem to stare back. And *Citizens of the Seas: Wondrous Creatures from the Census of Marine Life* (Plate 5) stages spectacular creatures, some within their stylized grids, others escaping them.[66] Such conceptions of sea life being welcomed as citizens will be discussed in chapter 3. For now, it is enough to consider that these portraits attempt to populate the abyss with beings worthy of regard and consideration.

Multispecies Perspectives and Ontological Musing

The question of whether or how speculations about species being can inspire volumetric environmental visions or maps of concern surfaces throughout this book. We could consider this question, following Bruno

Latour, as a "compositionist" matter, proceeding from the knowledge that the ocean is not "nature" in the modernist sense because it is not "always already assembled" but rather must be "composed" from "discontinuous pieces."[67] Latour's quest for the "Common World," however, even though it is "slowly composed instead of being taken for granted and *imposed* on all," suggests a unified transcendental perspective from which someone or something composes the arrangement of pieces.[68] While other entities and beings are represented as part of this composition, the belief in a resulting composition seems to be propped up by transcendent, immaterial composers. By contrast, Isabelle Stengers declares, in the volume *A World of Many Worlds*, "The global West is not a 'world' and recognizes no world. . . . A world destroying machine cannot fit with other worlds."[69] Noting that she herself had "evaded" or "tamed" the "question of other-than-human beings," she argues that an "ontological politics demands that we take seriously the existence and power of other-than-human beings," even when it threatens to destroy "all resources for thinking."[70] In critiquing the phobic Western denial of animism, Stengers suggests that other beings—as inert resources—are so fundamental to Western modes of thought that an ontological politics threatens thinking itself. What would it mean to think with—rather than upon, about, or against—more-than-human beings? Perhaps thought itself, as a rational, objective, disembodied, and depersonalized practice of mastering an externalized reality, needs to be muddled with more relational, embodied, and aesthetic vectors that attempt to think with a multitude of species and to imagine the fluid ontologies of the depths while relinquishing solid foundations of knowledge that would, say, capture and taxonomize living beings. However fraught and formidable, the philosophical project of imagining the perspectives of various creatures could invite thought that does not objectify the dazzling and precarious life in the sea. Several writers in this study, from William Beebe to Rachel Carson to John Wyndham to Nnedi Okorafor and Alexis Pauline Gumbs, engage in multispecies speculations, creating narratives from the perspectives of abyssal creatures, recognizing them as kin, or vertiginously spinning as they confront the impossibility of imagining the perceptions and lifeworlds of unfathomable beings.

As the impending horror of deep-sea mining threatens to destroy colossal benthic ecosystems, many scientists are not hopeful about the ability to prevent this massacre. Marine biologist Helen Scales quotes Daniel Jones of the British National Oceanography Centre: "Even if we found unicorns living on the seafloor," he says, "I don't think it would necessarily stop mining." Scales notes how easy it is to ignore the depths: "As soon as you stop thinking about it, the deep can so easily vanish out of mind—more so than that other great distant, realm, outer space."[71] However, she underscores the "invisible connections" that "lead far and wide from the deep sea, keeping balance in the atmosphere and climate, storing away and pouring our vital substances, all processes without which life on Earth would be unbearable or impossible." Emphatically, "every living thing needs the deep."[72] While her environmental passions drive the logical arguments and presentation of scientific data, she also includes vivid depictions of marvelous creatures, but her attempt to depict the horrors of seabed mining from the perspective of benthic and other sea creatures may be what is most compelling:

> Plumes of mining tailings would inject dust storms into their midst, including fine sediments that would hang suspended for years and get carried by ocean currents for hundreds of miles. Delicate animals of so many kinds—ctenophores and siphonophores, gossamer worms and bomber worms, larvaceans and jellyfish—would be smothered and dragged down by particles settling on them, their gills and delicate feeding apparatuses clogged so they can't breathe or eat. Dust clouds would substantially absorb blue light, selectively blocking the most common color the bioluminescent animals use to communicate. Their blinking lights and messages to lure and warn would be muted and erased in the murk.[73]

This terrifying scene, told from the third person but focalized through the perspectives of abyssal animals, is complemented by warnings that deep-sea mining could stir up toxins that would contaminate fisheries as well as churn up stores of carbon, thus exacerbating climate change. The book concludes, however, with a bleak impasse between the need for new energy systems and the contention that those systems require the metals found in the deep sea. This information is presented in a cold, factual, and distant manner, leaving readers feeling disconnected from the problem, with no orientation toward political activism, ethical practices, or ocean conservation. Ultimately, Scales does not help us develop a sense of "our own situatedness" that would illuminate our "patterns of consuming the world," as Marietta Radomska and Cecilia Åsberg put it, or an "ethical imagination" adequate for the Anthropocene.[74] A recent call for a ban on deep-sea mining explains how the "cultures across the Pacific" see the ocean as a "sacred space for creation, a provider, an ancestor," underscoring familial interdependence. This petition, "Indigenous Voices for a Ban on Deep Sea Mining," emanates not from the cold, disconnected ontoepistemology of scientific objectivity or anthropocentric and utilitarian "environmentalisms" but instead from kinship networks that extend to the seafloor: "For millennia our people have lived in a relationship with the natural world that is defined by respect, gratitude, responsibility, and love. Our genealogies, woven across space and time, connect us physically and spiritually to animals and plants from the highest mountains to the deepest ocean."[75] In itself, this is a potent, cohesive, and inspiring protest against mining. While many non-Indigenous scientists and environmentalists have stressed evolutionary models of kinship with ocean life, those origin stories often conclude without any ontological, ethical, or political significance, without any concern for marine life.[76] Moreover, Indigenous views can be encased within non-Indigenous frameworks—included in a manner that does not disrupt the dominant framing. The "Indigenous Voices" petition, signed by seventy-two Indigenous groups as of July 2024, appears as part of the Blue Climate Initiative website, which also features "Mineral and Genetic Resources" as one of its projects. The term "resource"—implying inert matter waiting to be used by extractivist capitalism and colonialism—resonates more with objectifying "development" regimes than with kinship and Indigenous sovereignty.[77] Once relatives are reduced to resources, respect and responsibility cannot flourish.

Because, as Scales suggests, it is easy to forget about the existence of deep-sea life, speculating about the perspectives of marine animals is a vital practice for creating imaginaries that expand environmental concern. Jakob von Uexküll contemplates the standpoint of different animals, proposing that even space and time are relative to each creature. For the deep-sea medusa, which moves in a constant rhythm, for example, "the same bell always tolls, and this controls the rhythm of life."[78] Michael Marder, in *Plant Thinking: A Philosophy of Vegetal Life,* contends that the "spatiality of all living beings—unmoored from objective determinations and emancipated from a global, disincarnated perspective that disavows its own perspectivalism—will require that a different sense of what is above and below, etc., be laboriously worked out from the standpoint of each particular life-form in question."[79] Laborious indeed! While it would be impossible for anyone to create such a map, imagining what it would entail renders transcendent global visions delusional. Imagining species perspectives, as a practice undertaken against a horizon of impossibility, may be an ethicopolitical incitement. Jonathan Balcombe, attempting to elicit empathy for "our underwater cousins" and convince readers that fishes "are individual beings whose lives have intrinsic value" and should be "included in our circle of moral concern," devotes chapters in his popular science book *What a Fish Knows* to what a fish "Perceives," "Feels," and "Thinks," how they socialize, and more. One chapter concludes by stressing the contextuality of intelligence, churning up conventional hierarchies that leave fish near the bottom: "When fishes outperform primates on a mental task, it is another reminder of how brain size, body size, presence of fur or scales, and evolutionary proximity to humans are wobbly criteria for gauging intelligence. They also illustrate the plurality and contextuality of intelligence, the fact that it is not one general property but rather a suite of abilities that may be expressed along different axes."[80] Along with Widder, we can marvel at an axis of intelligence signaled by the possibility of bioluminescent communication, heralded as the most ubiquitous mode of communication on the planet. Unlike the oft-forgotten fishes, cephalopods provoke not only philosophical speculation but fandom. Vilém Flusser and Louis Bec in *Vampyroteuthis Infernalis,* for example, their weird meditation on the vampire squid from hell, attempt to conjure the Dasein of this cephalopod, to "begin to see with its eyes and grasp with its tentacles: This attempt to cross from our world into its is, admittedly, a 'metaphorical' enterprise, but it is not 'transcendental.' We are not attempting to vault out of the world but to relocate into another's. Our concern is not with a 'theory' but with a 'fable,' with leaving the real world for a fabulous one."[81] What could be more fabulous than imagining a constellation of multispecies perspectives, even if each one is merely a tentative glimpse, a fleeting impression, a speculative foray?

Despite Emily Dickinson's musings, the brain is not "deeper than the sea"; the human brain cannot absorb it "as sponges buckets do" because there is no "it" there, no "sea" as such (whatever that "as such" could mean), but instead multitudes of interacting species, each with (or without) its own brain, as well as ecologies, substances, and forces that make marine animal studies, like the marine sciences, a formidable venture.[82] Dickinson's characteristic dashes—epistemological fits and starts that infuse the confident assertions with skeptical whimsy—may be the truest aspect of the poem. Cary Wolfe, in *Ecological Poetics, or Wallace Stevens's Birds,* draws on Jacques Derrida's *The Beast and the Sovereign,* systems theory, theoretical biology, and more to argue for a "*nonrepresentational* understanding of ecopoetics" in Stevens's poetry, noting that as contemporary biology asserts, "no organism has a representational relationship to its environment, in the sense of a neutral transparent access whose veracity and usefulness is calibrated to the degree of this neutrality and transparency."[83] Instead of seeking to represent an objective reality, freezing it into an accurate map of what is, we can understand, with Wolfe and Stevens, ecological space as "*virtual* space" because "any such space is populated by myriad wildly heterogenous life-forms that create their worlds, their environments, through the embodied enaction, unfolding dynamically and in real time, of their own self-referential modes of knowing and being, their own autopoesis."[84] Unlike conventional models of representation that still predominate in the humanities, this virtual space overflows with the knowing, doing, being and (self-)making of a multitude of more-than-human species. This is not less real but more so, as Wolfe notes "it's a wild, crisscrossing dance of an almost unimaginable heterogeneity of living beings, at different scales and at different temporalities, doing their own thing."[85] Wolfe develops this heterogeneity further with the term "jagged ontologies," concluding: "Paying serious attention to the question of 'the Animal' forces us to think more clearly and more rigorously about the biosphere in all its singularity and uniqueness in ways that reach far beyond the question of climate change and the Anthropocene. To make sense of any of these, we have to start with the realisation that what's needed here is not flat but ever more jagged ontologies."[86]

Contemplating heterogenous life-forms as they create their worlds would be vertiginous enough on land, but with a million or more species in the ocean (plus those that are as yet unknown), such contemplations warp into something akin to nitrogen narcosis. Irigaray's poetic ecofeminist deconstruction of Friedrich Nietzsche imagines a radically egalitarian oceanic multitude of beings: "The sea shines with a myriad eyes. And none is given any privilege. Even here and now she undoes all perspective."[87] While Irigaray subverts mythical and philosophical misogyny and anthropocentrism in this passage, by both critiquing and inhabiting a feminized sea, gender dualisms as well as the distance between human and sea creature disappear in Jorie Graham's poem "Deep Water Trawling," which conveys the violence of trawlers who smash the habitat and destroy "hundreds of species," at "2000 meters and more— / despite complete darkness that surrounds me— / despite my being in my place under strong pressure."[88] Graham graphically depicts the horror of being a deep-sea animal, with the "midwater nets like walls closing around us" and "the hammer" that "knocks the eyes out," while mentioning the pervasive anthropogenic harms to life in the sea: industrialized fishing that discards up to 90 percent of the catch, pesticides, dead zones, abandoned ghost nets that kill forever, and, ultimately, the end of the world.[89] Whose world? The poem answers that question by condemning human presumptions of ownership as wrong while, even more provocatively, posing the speaker as simultaneously or sequentially a deep-sea being and a human at the start: "am I human we don't know that."[90] Despite the bleak portrayal of anthropogenic ecocide, the speaker expresses an anachronistic and idealized sense of what the human could be: "Did you ever kill a fish. I was once but now I am / human. I have imagination. I want to love. I have self-interest. Things / are not me."[91] The speaker rejects objectification in a paradoxical, tentative, even contradictory manner, asserting that "things" are "not me" rather than speaking as a subject to say, more directly, "I am not a thing." Once the speaker becomes human, not fish, awkwardly, the being of the fish is unnamed: "I was once but now I am." This suggests both that the species is unknown to humans and that this mode of creaturely being was, in the past, a vital mode of being in a way that to be "but now" human is not. If imagination and love are predicated on self-interest, and if objectification of others shores up the self, then an ecological imagination of life in the ocean depths must be as disconcerting as this poem, which speaks as something that both is and is not human, calling the perpetrators of this invisible destruction to account while crushing the claims to innocence or disconnection that would deny the depth, breadth, and temporal scale of anthropogenic harm.

While we may marvel at visual images of deep-sea creatures, that aesthetic experience is ultimately meaningless without a commitment to the survival of marine life. We might note, bleakly, that the title of the Deep Sea Conservation Coalition's online video game is called "Game Over," with the subtitle "Is It Game Over for Residents of the Deep?"[92] Sue Reid, writing about the "sessile ones," the inhabitants of the deep-sea floor who will be devastated by mining, argues: "At a time when planetary environmental systems are in stress and decline, there is a vital place for imaginaries with which we might all navigate and transition. Thinking and imagining relationally and ecologically cultivates more sensitive interactions with ocean ecologies."[93] Envisioning the deep seas in a way that would matter ecologically demands not only that we envision the beings, lives, and worlds of animals in the depths but also that the invisible capitalist plunder of the open and deep seas becomes a matter of concern, a strong current of activist knowing that impels action in multiple domains. Such epistemological, ethical, and political work, undertaken by transcorporeal environmental subjects who find themselves immersed in networks of risk and responsibility, may seem to exist in another world, one far from the stunning images of abyssal life, which are portrayed as perfect specimens, dazzling aesthetic objects, or aliens from a distant realm, untouched by anthropogenic harms. It would, alas, be easy to slip into an abyss of cynicism and despair, not only because ocean ecologies face accelerating destruction but also because biodiversity and extinction have been overshadowed by an environmentalism concentrated almost exclusively on climate change and devoted to shoring up the lives of the most privileged peoples. Moreover, the very images of deep-sea creatures that inspire concern for the abyssal biome arrive through some of the same technologies used by industries that threaten ocean ecologies. Such troublesome circuits of mediation and knots of entanglement are complicated to navigate. Yet Radomska and Åsberg's call for a "low trophic" theory resonates: "How can we theorise in ways cognizant of our own patterns of consumption, potential violence, complexity and ecologies in which we as subjects, living beings, creators and knowledge producers are implicated?"[94] Such theories and practices require scientific and activist knowledges, which may be inspired by mediated aesthetic encounters with abyssal life.

Aesthetics as a Lure for Creaturely Speculation

Haraway charges that "these times called the Anthropocene are times of multispecies, including human, urgency: of great mass death and extinction; of onrushing disasters, whose unpredictable specificities are foolishly taken as unknowability itself; of refusing to know and to cultivate the capacity of response-ability; of refusing to be present in and to onrushing catastrophe in time; of unprecedented looking away."[95] While it is possible to find videos, photos, and illustrations of the destructiveness of deep-sea trawling, massive industrialized fishing ships, and the heaps of bycatch routinely killed by the fishing industry, for the most part, the scale of death and ecological decimation on the high seas would certainly qualify as an "unprecedented looking away." At the same time, however, the iconic "newly discovered" deep-sea creatures, weird and wonderful, appear in popular media. We might rewrite John Berger to say, "Everywhere ocean creatures disappear. In digital images and coffee table books they constitute a monument to their own disappearance."[96]

When we look at a photograph of a deep-sea animal, what relation can there be between that being—so distant, so strange, so alien—and the viewer? Kaja Silverman, in *World Spectators,* insists on the power of visual pleasure as well as the agency of what is seen. Critiquing the poststructuralist insistence on the primacy of language, she insists not only that "visual perception comes first" but that we look at "other creatures and things" "in response to their very precise solicitation to do so."[97] Drawing on Freud, she argues, "The pleasure principle can best be defined as the enabling force behind a particular kind of looking: the kind of looking which is *creative* of beauty or preciousness. It is the impetus driving us to find visual gratification in perceptions that only imperfectly replicate our memories, and—in so doing—*to ennoble* ever new creatures and things. It is that to which we owe our capacity to affirm the phenomenal multiplicity of our earthly habitus: to become world spectators."[98] *World Spectators,* published in 2000, telegraphs the need for the nonhuman turn. In the previous quote, for example, Silverman states that it is human viewers who "create" the beauty of these creatures and "ennoble" them. In her formulation, "the world," seemingly a single entity outside the human, seduces us. She writes, "The world does not simply give itself to be seen; it gives itself to be loved."[99] This conceit, while it turns on the world's agency in giving of itself, concludes with a crushing anthropocentric embrace. Silverman insists, "It is we alone who provide the light by means of which creatures and things appear."[100] Reading this in the wake of critical posthumanism, animal studies, multispecies studies, and the environmental humanities underscores its blatant human exceptionalism. Yet for the matter at hand, there is some truth in the idea that "it is we alone who provide the light by means of which creatures and things appear," in terms of the capture and dissemination of these highly mediated images. Moreover, the way these images circulate corresponds to Silverman's contention that "the phenomenal forms of the world invite us to make them part of our singular language of desire—to make them components of the rhetoric through which 'we care.'"[101] Beauty, desire, pleasure, and caring are imbricated. In Silverman's more recent volume, *The Miracle of Analogy,* which ignores posthumanism and environmental theory, she nonetheless casts photography as a leveling, disanthropocentric medium, calling it an "ontological calling card" that "helps us to see that each of us is a node in a vast constellation of analogies." This vast constellation of relations seems rather ecological, especially because "authorless and untranscendable similarities . . . structure Being," giving "everything the same ontological weight."[102] Chapter 3 discusses the stunning oversize photographic collection of deep-sea life by Claire Nouvian. Nouvian's compositions pose even the faceless abyssal animals in a portrait-like manner, each creature holding "the same ontological weight," presenting viewers with an enticement to reflect on their being without hierarchal scales of high and low. At the far reaches of Silverman's theory is the contention that "it is only through this interlocking that we ourselves exist. Two is the smallest unit of being."[103] It is breathtaking to consider coming into an intersubjective, multispecies mode of being—even as a wispy, billowy moment—through such highly mediated relations and "ontological calling cards." Yet if we remove ourselves from the scene, we could also ponder an aesthetics of the abyss that does not require humans as audience or composer, asking with the deep-sea scientist from J. M. Ledgard's novel *Submergence,* "Did the abyss sing of itself?"[104]

Such a poetic question is alluring. It is not possible to ever know what it is like to be a bat (with Thomas Nagel),[105] or a whale (with Burnett), or a hatchetfish (with Beebe), but the practice of creaturely speculation may nonetheless be vital for animal ethics and environmentalism. Steve Mentz, in *At the Bottom of Shakespeare's Ocean,* suggests that the ocean poses a "basic challenge: to know an ungraspable thing." He states that "Shakespeare's plays write the sea as opaque, inhospitable, and alluring, a dynamic reservoir of estrangement and enchantment."[106] The sea, both enchanting and opaque, entices us to contemplate its being and beings. Against the overwhelming scale of the imagined entity of the ocean, distinct species and organisms appear, inviting a paradoxical intimacy that propels the pleasure of wonder and the commitment to concern. We can tack between aesthetic pleasure that seems like a sensual, intimate encounter with another being and the provocation posed by the realization that the image conveys something of the being depicted, yet little understanding of that creature. Aesthetic pleasure sparks speculation—not only about creaturely lives but about species' habitats and precarious futures. María Puig de la Bellacasa writes in *Matters of Care: Speculative Ethics in More Than Human Worlds:* "That things could be different is the impulse of speculative thinking," adding that the term "speculative" refers to a "mode of thought committed to foster visions of other worlds possible," a "political imagination of the possible."[107] To imagine—as part of an environmental politics, or more specifically an oceanic environmentalism—is to foster a sense of concern that provokes public support, policies, laws, and a multitude of everyday practices.

As a feminist, environmentalist, posthumanist, and cultural studies scholar rather than a philosopher, I steer *The Abyss Stares Back* away from rarified debates about aesthetics in favor of attending to flagrant accounts of the beautiful, the dazzling, the surreal, the weird, the unfathomable, and the alien as they circulate through science, art, and popular culture. Rather than precisely parsing the differences between the sensual, the sublime, the emotional, and even the cognitive, I hover with the aesthetic as a mode where these responses swirl together. While humanities scholarship values definitional precision, in this instance, taxonomizing or narrowing the concept of the aesthetic would be counterproductive to the posthumanist work I intend the concept to do. A capacious, inclusive, and potent creaturely aesthetic muddles scientific objectivity with emotion, ungrounds gendered dualisms of thought and feeling, and invokes a sense of multispecies encounters that are staged through networks of scientific and artistic capture. To label and divide different modes of the aesthetic at the outset would be to resist the siren song of the aesthetic pull of abyssal life, which seduces us with promises of mediated intimacy, sparks curiosity and awe for the singularity of a multitude of aquatic creatures, and, it can be hoped, provokes more expansive, volumetric, animal-oriented terrains of environmental concern. Perhaps deep-sea creatures can propel a shift from the ontoethicoepistemological of new materialist theory[108] to an even more vast and murky place where the aesthetic—the predominant mode in which deep-sea life is encountered—not only claims a place alongside the ontological, epistemological, and ethical but also slips in dodgy inhuman modes of the political, infusing all of these categories with pleasure, sensuality, relationality, affirmation, and awe.

◈ 1 ◈

Animating Surreal Creatures a Half Mile Down

William Beebe and Else Bostelmann

> You exclaim something bromidic which sounds like Marvellous! Great! Wonderful! then relapse futilely into silence and look helplessly into the distance where the emerald waves still break and the palms wave as if fairyland had not intervened in your life since you saw them last.
>
> —William Beebe, *Beneath Tropic Seas*

> Too much fairyland! Too much eerie beauty!
>
> —*Killers of the Sea*

The writings of William Beebe and the surreal paintings of Else Bostelmann invite us to muse about deep-sea species—creatures so strange and so beautiful that they unsettle assumptions about life itself, provoking us to question the nature of science and its relations to the emotional, the aesthetic, and the marvelous. If one glimpsed something billowy and gelatinous, something unknown, while peeking out the window of a submersible a half mile down in the ocean and felt a sudden intake of breath, how, then, could that embodied aesthetic response flow into a practice that would be deemed properly scientific? Does the encounter with a radically different being elicit a response that is at once cognitive, emotional, embodied, and aesthetic, revealing the fluid borders between these established domains? While the Western practices and narratives of discovery, taxonomy, and natural history are saturated with colonialism, is it possible for submersive encounters to corrode the armor of mastery? Can the eerie beauty of the deep sea provoke less terrestrial, less anthropocentric, less rigidly disciplined responses? And what happens to science when it descends into a fairyland?

A Brutal Prelude

Aesthetic captivation by, sympathy for, and speculation about deep-sea creatures undermines the dominant Western sense of the ocean as an empty surface,[1] the colonialist, extractivist, and capitalist assumption that marine life is there for the taking, and the brutal exploits of white man making. While this chapter focuses on the writings of William Beebe and the paintings of Else Bostelmann, particularly those that depict deep-sea animals in the 1930s, I would like to begin with a grisly contrast to what follows, namely a rather violent rejection of the fairyland and eerie beauty of coral reefs, which suggests the potential for aesthetic captivation to disorient narratives of mastery.

In "Science vs. Showmanship on the Silent Screen," Greg Mittman writes that in the 1920s, "Hollywood spared no expense to procure the macabre": "killing game and predators killing prey were staples of the travelogue-expedition films of the 1920s, scenes that audiences had come to expect."[2] Following this tradition, the next decade produced the awkwardly violent and inadvertently campy movie *Killers of the Sea* (1937), a documentary that follows Captain Wallace Caswell Jr., who dives off a small boat to battle "the killers," "the destroyers," "the monsters," and the "tyrants of the Gulf," who supposedly rob fishermen of their catch and ruin their nets. The opening text, white letters rolling against an ocean background, explains: "He is the G-man of the sea, stalking its deadliest Killers, armed with quick brain and steel-like muscle, it is man against Monster in his native element." Each time a "killer" comes near, Caswell strips down to dive into the ocean to wrestle the animal. (As the male narrator gleefully repeats, "Off with those pants!") The film documents his heroic exploits with shots of newspaper headlines and punctuates the drama with outbursts such as, "stab, stab stab . . . killing the killer!" Although the narrator warns, "When he comes to grips with a killer—then he'd better be a man of steel," the puny captain does not actually sport an impressive muscular physique. Many of the animals that he wrestles, stabs, and kills—including a sea turtle and a hammerhead shark—in horrifically slow, awkward, and gruesome encounters—are not much bigger than he is, including the "whale," "Moby-Dick's cousin," who is actually a diminutive dolphin. While the captain is not, despite the acclaim, a muscular man of steel, he is certainly white, as the racist and ableist shots of a black man called Evolution, who is missing half a leg and jumps in after a small fish he caught, not so subtly underscore. When another crew member who is Black wrestles a dolphin, he receives no praise for his heroism, intelligence, or physique. Instead, the narrator grandly presumes that the events dwarf the man's imagination: "What colored man ever dreamed of having a leviathan in his boat?" The racism is hardly surprising, given the well-documented articulation of white American masculinity with "hard bodies," aggression, and violence, as well as Teddy Roosevelt's conception of the wilderness as the site for white supremacist man making.[3] Yet the ridiculous insult is still telling, suggesting as it does that white men's aspirations naturally include having a leviathan in their boat—or, for that matter, all other living creatures under their control. The white possessive steers the ship.[4] Oddly, the underwater scenes are overlaid with the exclamation, "Too much fairyland! Too much eerie beauty!" as the small fish and other living creatures interfere with the captain's killing spree. "The diver groping his dim underwater way is impeded by the fish that flock about him. . . . In this world of the unhuman he doesn't seem to frighten the finny folk at all." Ironically, the murky underwater shot fails to capture the beauty of colorful creatures. The film concludes less than triumphantly, as the captain is cut by a sawfish and hospitalized. Yet white masculinity marches on, seemingly unscathed, secured by the "comic" inclusion of Evolution, as well as by the brutality waged on sea creatures and the rejection of opulent undersea beauty, which, one suspects, shimmers with a gay and/or feminine, flamboyant decadence.

Ten years before this film was released, William Beebe began his book, *Beneath Tropic Seas* (1928), with a chapter called "Brothering Fish," which presents an ecstatic account of helmet diving in the Haitian coral reefs. It should be noted that "brothering" Haitians was not on the agenda, opening a direct line from Beebe to the fish. Zakiyyah Iman Jackson's astute argument that "discourses on nonhuman animals and animalized humans are forged through each other; they reflect and refract each other for the purposes of producing an idealized and teleological conception of 'the human'"[5] illuminates the structure of relations between the "ideal" human as a mobile intrepid adventurer, the local Black and colonized peoples, and the fish. Beebe's passion for ocean life must be understood within the racial and colonial formations that subtend the relations between the human and marine species. As Claire Voon argues, referring to Beebe's expeditions with the Department of Tropical Research (DTR), "The ugly reality is that DTR set up its first field station in British Guiana, and its research relied heavily on the established systems of colonial control. The team hired locals as guides, cooks, and even specimen preparers, and they viewed them as no more than laborers."[6] Voon identifies a "small clue to the researchers' disregard for those who resided on these lands long before their arrival" in a map drawn by Helen Damrosch Tee-Van, an artist on Beebe's team, depicting an "imagined version of Bermudas' Nonsuch Island," which, along with "Tee-Van's illustrations of animals, nature, and the DTR members going on their daily agendas are racist caricatures of three individuals."[7] Ann Elias's study, *Coral Empire: Underwater Oceans, Colonial Tropics, Visual Modernity,* focusing on the underwater filmmaking and photography of Ernest Williamson in the Bahamas and Frank Hurley in Australia in the 1920s, includes images that reveal "how black and indigenous peoples were either pushed to the background or turned into spectacles" by Williamson and Hurley.[8] My archival research on Beebe's deep-sea descents and Bostelmann's scientific art has not yielded anything comparable to what Voon and Elias describe, but of course Beebe's exploits are undeniably part of the fabric of racist imperialism. Beebe dedicates his account of the bathysphere dives, *Half Mile Down,* to racist eugenicist Madison Grant. Moreover, the DTR was part of the New York Zoological Society at the Bronx Zoo, where not only Grant but also Harry Fairfield Osborn Sr. were founders, and where, most notoriously, Ota Benga was "displayed."[9] Many of the essays and much of the art produced by Beebe, Bostelmann, and the rest of the DTR members were published in *National Geographic,* the premier purveyor of coffee-table colonialism. Any type of animal ethics, environmentalism, or posthumanism that we may identify in this archive has been articulated within and as part of a sea of white supremacy and colonialism. The scientific gaze feels no need to justify its global scope. The science and art that Beebe's team practiced were unabashedly extractive, taking not only animal specimens from the sites but also taking the knowledges of Indigenous and other local cultures, and relying on residents for labor. As Richard Drayton explains in an interview with Katherine McLeod, referencing Beebe's scientific publications from fieldwork in British Guiana, "At no point was he likely to give credit to all of the local informants who would have guided him towards the phenomena he came to understand." McLeod adds that Beebe spoke for the people "in the same way he speaks of plants and animals . . . coopting . . . their knowledge as his own."[10] Science journalist Joanna Klein assesses the legacy of the DTR:

> Although they helped lay foundations for modern conservation, field biology and animal behavior studies, they were riding the coattails of colonialism and industrialism, often funded by America's economic elite as it sought to control natural resources in the Caribbean and South America. Indentured workers from India, descendants of slaves from Africa, and Native American groups were critical to the DTR's work, but merely mentioned in some texts.[11]

Moreover, it is likely that Beebe imagined his readers, as well as himself, as "universal" white people, with an appetite for adventurous discoveries underwritten by colonial imperialism and racial exclusions. While I do not focus on racist or colonial visions in this chapter, they not only haunt the horizons but also permeate the imaginaries herein. Katherine McKittrick's plea to "be uncomfortably satisfied with the unmeasurability of black life and to engender interhuman relationalities" and to "honor a sense of black place," although pertaining to other topics, marks crucial absences here.[12] Recent scholarship on Black being and aliveness beckons toward speculative histories, poetics, and theories beyond the archive of this chapter.[13]

Notwithstanding the colonialist histories that Beebe and the DTR team perpetuated, Beebe's aestheticization of sea life counters the (unintentionally) campy *Killers of the Sea,* as well as the 1914 films of Williamson, who, Elias argues, projected "racial difference onto animals of the sea" and projected sharks "onto the racialized other."[14] Elias writes that Williamson's films

> aimed to immerse audiences in fearful fantasies, portray the racial Other as a natural part of primitive wilderness, stimulate-imaginations with morbid fears of monsters, suggest that the fore-

> boding underwater harbors bitter struggles for man and beast and project the idea that because the deep sea is a wilderness, more fearful than the jungles of the land, it was a region that must be conquered.[15]

While Williamson pits the lone hero against the monstrous other, in *Beneath Tropic Seas: A Record of Diving among the Coral Reefs of Haiti* (1928), Beebe greets fish as family. This is preferable from the standpoint of animal ethics and environmentalism, yet Beebe's sanguine encounters with the fish are predicated on the convenient erasure of the human inhabitants of the region. Beebe narrates an exuberant response to his encounter with ocean life in the second person, calling the reader to experience it with him: "Your attention swings from wonders to marvels and back again. You begin to say things to yourself, gasps of surprise, inarticulate sounds of awe."[16] The second-person "you" collapses the speaker and the audience into one person, a unity that belies the silent expulsion of other persons from the scene, clearing the waters for amiable encounters between the white explorer (along with his presumed white readers) and the fish themselves. Given the racist ideologies associating blackness with a denigrated sense of animality, I suspect that Beebe consciously or unconsciously erased Black people from the scene, in part to avoid tainting his idyllic vision of marine life. Ironically, however, segregating the fairyland from Black people drenches the scene in racism—less overtly than in *Killers of the Sea,* yet more insidiously. It is crucial to underscore this subtle yet deeply rooted racist structure—a structure that persists today in articulations of wilderness, adventure culture, and ecotourism as presumptively white. While I cannot resolve this trouble here, I do not want to minimize it. Rather, I suggest how it complicates and compromises the content of much of this chapter.

Beebe does not foster a sense of brotherhood with the Haitians, but neither does he occupy a position of mastery toward the underwater life that astonishes and overwhelms him. He is not there to kill or to (homo)phobically flee from flamboyant beauty. The fairyland of the sea, which, Beebe writes, evokes "gasps of surprise and inarticulate sounds of awe," has the power to "intervene" "in your life," rendering human modes of linguistic capture futile. Margaret Cohen applies Jonathan Lamb's thoughts on the je ne sais quoi to this and other passages in *Beneath Tropic Seas,* noting that it is "a figure that occurs when observers reach the limits of their prior knowledge of the world," which "explains why navy divers, naturalists, and engineers would turn to fantasy images and allusions to express a new planetary frontier."[17] Despite Beebe's inarticulate awe, the "curious" and "friendly" fish manage to communicate, greeting Beebe, with "Oh! Oh! Brother! Brother! Oh! Oh!," seeming to be as excited as Beebe is by this encounter, embracing him as family. As the experiential aesthetic of the dive undoes the diver, he implores readers to join him in this ecstasy: "All I ask of each reader is this,—Don't die without having borrowed, stolen, purchased, or made a helmet of sorts, to glimpse for yourself this new world."[18] When Beebe descends into the deep seas in the 1930s, he will be even more frustrated with his inability to verbally represent the creatures he glimpses, drawing on the talents of the artists as supplements. *Killers of the Sea,* released a few years after William Beebe and Otis Barton made their historic descents a "half mile" down into the ocean in a bathysphere, underscores the connection between violence and heteronormative white masculinity, suggesting the potential for a surreal aesthetics—the eerie beauty of the fairyland at the bottom of the sea—to entice, entangle, and elude the sort of capture that would stabilize the domains of scientists or adventurers. Blue humanities scholars and other lovers of ocean life can revel in the refrain "Too much fairyland! Too much eerie beauty!" while marking Beebe's constitutive exclusions.

William Beebe, Natural History Traditions, and the Dualisms of Modern Science

Beebe's illustrious life and work resist compression. Characterizations of Beebe and assessments of his scientific, literary, conservationist, and popular legacy swerve. The official William Beebe website states simply, "He is most famous for his world record descent in the Bathysphere with Otis Barton on August 15, 1934. Will also wrote popular books about his exploring experiences working for the New York Zoological Park (Bronx Zoo)."[19] Robert D. Ballard, in *The Eternal Darkness: A Personal History of Deep Sea Exploration,* introduces him as a "fascinating mixture of scientist, poet, showman, and explorer."[20] Natascha Adamowsky presents Beebe as a "natural historian, avid diver," and the director of the DTR, noting that he possessed "a singular knack for public relations, and he was a genius when it came to raising money for expeditions."[21] Historian Eric Paul Roorda describes him as a "tabloid mash-up of Captain Nemo and Indiana Jones," noting that his "standing in the history of science seems slippery" but that "his greatness as a writer of unequaled range has never been recognized."[22] Gary Kroll, in *America's Ocean Wilderness: A Cultural History of Twentieth-Century Exploration,* offers a rather harsh assessment, calling him the "ad man of natural history" who used the sublime to distance himself from the "odious stuntlike nature of heroic adventure."[23] In a brief mention of Beebe, D. Graham Burnett compares Gifford Pinchot's South Seas expedition to "celebrity nature grandstanding in the best tradition of Teddy Roosevelt and William Beebe."[24] Carol Grant Gould concludes her biography of Beebe by underscoring "the driving force of his unquenchable enthusiasm for the natural world," noting that his "exhilaration in its wonders inspired a generation of ardent nature lovers."[25] Any assessment of Beebe in terms of the blue humanities should underscore that he popularized ocean exploration well before Jacques Cousteau. Such achievements were received within, and no doubt benefited from, masculinist discourses of white colonial "discovery." A typed anonymous statement for the 1953 Roosevelt Medal, accompanied by a medallion, ends by describing Beebe as "a Columbus of the sea-deeps, daring the black unknown that no human eye before him has penetrated, and returning with tales of creatures, fabulous and fantastic beyond anything that fevered fancy could project."[26] The references to Columbus and to "penetrating" the "black unknown" call for Calvin L. Warren's critique of a black void, as well as the Black feminist speculative theory of Alexis Pauline Gumbs, who writes in *M Archive: After the End of the World,* "they had this thing about the darkness. the bottom of the ocean, outer space. they were afraid of it they wanted to penetrate it."[27]

Racist and sexist receptions of Beebe's exploration notwithstanding, Beebe did produce "fabulous and fantastic" tales of ocean creatures that captured the attention of his public and inspired the development of marine conservation. It would be an understatement to say that Beebe was a prolific writer; he left a massive amount of published and unpublished, strictly scientific as well as more literary, popular works. In terms of environmental studies, it is notable that Beebe undertook a popular form of nature and adventure writing that did not divide science from literature or aesthetic appreciation from scientific pursuit. Beebe is included in the *Dictionary of Literary Biography,* volume 275, *Twentieth-Century American Nature Writers: Prose,* his writing categorized within the "travel writing and natural history" traditions of the "nature writing genre."[28] Kroll states that while *Half Mile Down* "received mixed reviews," Rachel Carson "informed Beebe that she had read the text four times."[29] Beebe remains a compelling figure in the arts and popular culture. In 2011, the Hull Philharmonic Orchestra commissioned Nigel Morgan to create "Sounding the Deep," a work for bass voice and symphony orchestra, with its libretto based on *Half Mile Down.*[30] At least four children's books feature Beebe and the bathysphere dives.[31] Kent E. Poltisch, author of thrillers, self-published a sensationalist historical novel entitled *Beebe and Bostelmann.* And Brad Fox published a beautiful work of creative nonfiction, "blurring the line between poetry and research," as the jacket blurb states, entitled *The Bathysphere Book: Effects of the Luminous Ocean Depths.*[32] Most notably, the exhibit at the Drawing Center, in New York, organized by Mark Dion, Katherine McLeod, and Madeleine Thompson, *Exploratory Works: Drawings from the Department of Tropical Research Field Expeditions,* in 2017 sparked renewed attention to William Beebe's research team. The exhibit featured artifacts from the field, a wealth of photographs, drawings, paintings, and short films, and a fantastic exhibit catalog with essays and full-color illustrations.[33] Dion, McLeod, and Thompson, in their introduction to the collection, argue that the DTR staff developed "new visual and written languages adapted to their immersive approach to studying the environment," adding that these "inventions in writing and drawing utilized what could be called a visionary poetics of ecological imaginations: the passages of writing and expressive methods of scientific illustration transport the imagination and act as jumping off points for thinking about relationships in nature, not as endpoints for displaying solidified facts."[34] Certainly Beebe and his staff invented a visionary poetics that was not subordinated to scientific facts; I would extend their argument to say that their verbal and visual depictions of deep-sea life in particular dramatize aesthetic encounters in such a way as to animate the dead specimen and lure readers or viewers into speculating about creaturely lives within the unimaginable depths. Just as art is not merely a tool for the use of science, the animal is not solely a specimen. A fluid spectrum of beings, encounters, and modes of knowing and speculating erodes the dualisms of modern science and its deadening gaze. The surreal style of Bostelmann's paintings, as I discuss below, self-reflexively underscores the intermeshed processes of art and science, while Beebe's reflections animate the specimen through narratives and philosophical musings.

Assessments of Beebe's scientific legacy are mixed, especially when it comes to his deep-sea explorations. Indeed, the reception of Beebe's scientific, literary, and popular achievements reveal a mottled, shifting, and fraught map of scientific parameters. Susan Schlee's chapter on "American Oceanography in the Early Twentieth Century," from her 1975 book *History of Oceanography,* includes zoology, but it does not mention Beebe. Marine scientist and National Oceanic and Atmospheric Administration official C. P. Idyll's 1976 *Abyss: The Deep Sea and the Creatures that Live in It,* includes at least twenty-one mentions of Beebe, many of which are long quotes from Beebe's descriptions of specific creatures.[35] Legendary marine scientist and conservationist Sylvia Earle, inspired by Beebe as a child, narrates how in the 1980s she "packed along her treasured copy of William Beebe's *Half Mile Down*" while supervising a descent, scrutinizing what Beebe had found "half a century earlier in the very depth."[36] Some critique Beebe's bathysphere descents but credit other aspects of his marine science. Ballard notes the controversy over *Half Mile Down,* explaining that one of the reviewers, Carl Hubbs, a scientist, "indignantly charged that Beebe had no right 'to describe and assign generic and species names for animals faintly seen through the bathysphere windows,'" and John Nichols, "a curator at the American Museum of Natural History, hinted that *Half Mile Down* should be classified as fiction rather than fact, because Beebe wrote in 'dramatic fashion rather than meticulous.'"[37] Ballard concludes, however, by noting that Beebe was a "pioneer of ocean ecology—largely because of his systematic sampling, his many indisputable specimens and his ability to inspire others to follow him into the abyss."[38] William J. Broad explains how Beebe and his colleagues undertook a thorough study of the seas off the coast of Bermuda between 1929 and 1937, when they "caught more than 115,000 animals representing 220 species of deep life, many of them new to science." Broad notes that "his work represented some of the most comprehensive and methodical sampling of one part of the ocean that had ever been done."[39] Tony Koslow, however, concludes his book *The Silent Deep: The Discovery, Ecology, and Conservation of the Deep Seas* by calling Beebe's thirty-five bathysphere dives scientific failures, noting that "without more concrete evidence—photographs or specimens—considerable skepticism accrued about Beebe's more fantastical-seeming sightings, and his dives . . . had more influence on the popular imagination than on oceanographic science."[40] Wolf H. Berger, in the textbook *Ocean: Reflections on a Century of Exploration,* credits the bathysphere dives with making the public aware of "a new world never seen before," noting that Beebe "caught glimpses of unnamed living things at the edge of his range of vision."[41] However, Berger insists in a footnote that ichthyologist Carl L. Hubbs was right to criticize Beebe in his 1935 review of *Half Mile Down* for naming creatures "faintly seen through the bathysphere window," underscoring the need for "proper procedure when naming organisms."[42] However, Adamowsky concludes that "all criticism notwithstanding, Beebe counted—and counts—as an important figure who identified many new species." Moreover, "Bostelmann's pictures are also considered important scientific illustrations to this day."[43]

Entitling the section on Beebe's bathysphere dives "Abysmal Research," Kroll charges that the descents were scientifically suspect because they could not produce specimens, arguing—although admitting, ironically, that he himself has "no direct evidence"—that "Beebe invoked the sublime to mitigate the unpalatably sensational nature of his descents."[44] Kroll quotes from Beebe's description of the bathysphere dives, which shifts from saying that he "cared nothing for any man-made measurements of depth—my only goal was to get beyond solar light" to a description of how they were "seeing creatures which had evolved in the blackness of a blue midnight for the past one hundred million years or so, ever since the second day of creation when oceans were born."[45] Kroll interprets this passage: "Beebe's concern that the bathyspheric dives be portrayed as a thoroughly scientific endeavor—in no way a spectacular stunt—gives way to sublime thoughts on the creation of all life."[46] Kroll argues that Beebe legitimized the bathysphere dives by his "invocation of the sublime," an "emotional aesthetic that is felt when one experiences a spectacle of unfathomable and ineffable awe," often "coupled with a certain failure of language."[47]

He concludes with a confusing set of assessments—that Beebe was neither a conservationist nor environmentalist, yet neither was he a utilitarian, but that he saw nature as having an "inherent aesthetic and spiritual value," and in the end, Beebe "made the ocean something worth protecting."[48]

Rather than distinguishing what is properly and laudably scientific, or rendering a verdict on Beebe's deep-sea scientific achievements, I would like to consider how Beebe's aesthetic responses, conveyed in his affecting prose, become something to be disciplined, cordoned off, or explained away. Just as certain scientific practices require dead specimens, so too do they seem to require the deadened emotions and ethics of scientists in the name of objectivity. Beebe's dramatic and self-reflective narratives foreground epistemological and philosophical quandaries, placing scientific practice within a more capacious intellectual domain.[49] Whereas Kroll sees Beebe's invocations of the sublime as a strategic way to cover up scientific failure or scrub the taint of sensationalism, I think Beebe experienced his descent into the deep sea, as well as the sight of strange and spectacular creatures through the bathysphere window, as ineffable, something beyond what could be contained within scientific and narrative captures, something that unmoored him. This unmooring can be understood in posthumanist and environmental terms as the aesthetic, emotional, and embodied sense that deep-sea life cannot be readily contained within the frameworks of conventional Western science. At the turn of the twenty-first century, as I discuss in the introduction, prominent deep-sea biologists such as Edith Widder pen unabashed accounts of marveling at deep-sea life, while Cindy Lee Van Dover, in *Deep-Ocean Journeys: Discovering New Life at the Bottom of the Sea,* cites a paragraph from *Half Mile Down* in which Beebe writes of being "inarticulate with amazement and with a deep realization of the marvel of what he has seen," adding that she follows "modestly in the footsteps of William Beebe."[50] Widder, Van Dover, and Sylvia Earle ("her deepness") marvel at abyssal and other marine life, popularizing the science, evoking concern for the species, and promoting a more expansive terrain of environmental concern through aesthetic currents.

Like later psychonauts who experimented with mind-altering drugs, Beebe attempts to convey his own unmooring through his inability to express what he sees and feels. Margaret Cohen's insightful interpretations of Beebe note instances in which "avant-garde art melded with a drug trip," as Beebe mentions "opium dream[s]" and "modernist or futurist designs" while framing his narrative with "gothic suspense."[51] Beebe's performance of what Cohen terms the je ne sais quoi contests the narrow bounds of scientific endeavor by placing the descent in a more capacious space where science, philosophy, and the aesthetic intermingle, a space where Beebe resists the constitution of the modern.[52] Alfred North Whitehead, in *Adventures of Ideas,* published in 1933 but based on lectures from 1926 to 1930, claims that "Science and Philosophy mutually criticize each other, and provide imaginative material for the other."[53] The je ne sais quoi that Margaret Cohen finds in Beebe and other writings by undersea explorers[54] resonates with Whitehead's insistence that beauty is truly novel and thus provokes a response beyond words: "Beauty is a discovery and not a recapitulation. The Truth that for such extremity of Beauty is wanted is that truth-relation whereby Appearance summons up new resources of feeling from the depths of Reality. It is a Truth of feeling, and not a Truth of verbalization. The Relata in Reality must lie below the stale presuppositions of verbal thought. The Truth of supreme Beauty lies beyond the dictionary meaning of words."[55] Rather than discipline Beebe's impressions and expressions into science proper, we could consider his discoveries, from within the depths of "Reality," to possess not only great beauty but also the ability to dispossess scientists from a sense of mastery, transcendence, and the blinders of mundane life. Whereas conventional scientific epistemologies divide the subject from the object, seeking to establish objectivity,[56] Whitehead insists that the "extremity of Beauty" provokes "new resources of feeling" that come not from the individual but from the very "depths of Reality." Whitehead's thinking can recontextualize Beebe's struggles to proceed as a naturalist undertaking adventurous expeditions as well as a humanist-scientist embarking on an "adventure of ideas." Such an adventure can veer away from the deadening objectification of the specimen, as the scientific practices of capture give way to being captivated. Being captivated, feeling unmoored, being overcome with wonder, having the solid ground fall away—all can provoke an ethical epistemology that recognizes the unknowable being of other species and their unfathomable lifeworlds. Although aesthetic desire can drive the objectification and exploitation of other species, it can also shift from a negative anthropocentrism to what Rosi Braidotti calls "positive passions," an "ethics of joy," and an "affirmation of one's interconnections to others in their multiplicity."[57]

Scientific Capture and Aesthetic Captivation

While scientific captures can be as immaterial as abstract data wrought by technological mediation, the evidentiary value of the animal specimen within natural history, taxonomy, and modern science cannot be overlooked. Collectors killed, purchased, or otherwise amassed heaps of dead animals captured in the field, seeking novel discoveries, for fame, science, nationalism, imperialism, and sheer rapacity. From a twenty-first-century perspective, when accelerating extinctions are painfully evident, the numbers of animals killed by those Richard Conniff calls "species seekers" from the mid-eighteenth century through the nineteenth century is horrific to consider. Walter Rothschild, for example, born into a British banking empire in 1868, kept "400 collectors actively working in all corners of the Earth" for "much of his adult life," racking up 2¼ million butterflies and moths, 144 giant tortoises, and many more animal bodies. He and his collaborators described "5000 new species."[58] Introducing Rothschild, Conniff explains that while early explorers "had set out to dazzle the world with heroic tales and strange new species," the new species seekers "wanted to have complete material for careful revisions of entire taxonomic groups." In short, "industrial scale natural history made for better science."[59] The scale of the slaughter underscores the deadly eye of scientific vision, reminding us that the desire to discover new species and construct taxonomies has contributed to the death of a multitude of animals—even to the point, in some cases, presumably, of pushing them over the edge into extinction. Local people, including Indigenous people, who relied on those species not only for food but as integral parts of their culture and, relationally, as kin or neighbors, were harmed, as Western scientific "progress" has been intertwined with racism and colonialism.

Criticism of Beebe's bathysphere dives, as discussed above, centers on the fact that Beebe attempted to describe what he saw from the submersible's small window, without the ability to capture specimens of what he glimpsed. Depending on the species, however, he may have already seen specimens captured by other means in that geographical area and depth. The bathysphere descents, however, enabled him to glimpse the creatures alive, in their own worlds, not writhing or pulverized in nets or trawls, or surrendering their secrets during dissection. Both ecological and ethological knowledge depends on in situ observation because the dead specimen can only say so much. Underscoring the value of these glimpses, Beebe notes, regarding one type of deep-sea fish, "Of all the many thousands of these fish which I have netted, I never saw one alive until now." Moreover, whereas the lanternfish caught in the nets would have "only a half dozen scales left," living fish in their realm "were ablaze with their full armor of iridescence," even presenting "the flash of their light organs."[60] At seven hundred and eight hundred feet, "a human being was permitted for the first time the sight of living, silver hatchet-fish, heliographing their silver sides."[61] Throughout *Half Mile Down,* Beebe justifies his bathysphere observations by critiquing "the inadequacy of our modern methods of trawling," noting the many discrepancies between the sort of species that trawling apprehended from various depths and the species observed from within the bathysphere at those depths, explaining that having seen the creatures "dart and twist and turn, their lights passing, crossing, and recrossing in bewildering mazes," it is no wonder that they manage to evade capture in the nets.[62] This vibrant scene might recall Cary Wolfe's vision, discussed in the introduction, of "a wild, crisscrossing dance of an almost unimaginable heterogeneity of living beings, at different scales and at different temporalities, doing their own thing."[63] Replying to the accusation that the dives were a mere stunt, Beebe argues that "all the trouble and cost and risk were repaid many fold" because the dead specimens in the nets do not reveal "their colors and their absence of colors, their activities and modes of swimming and clear evidence of their sociability or solitary habits."[64] Beebe conveys his breathtaking encounters with creatures in the depths as themselves invaluable, even if, or precisely because, they are beyond words: "My inarticulateness and over-enthusiastic utterances may well be excused on the grounds of sheer astonishment at the unexpected richness of display."[65]

The evaluation of the specimen as inadequate, in terms of scientific observation, aesthetics, and cross-species encounters, perhaps motivated Beebe to imaginatively reanimate the dying and dead animals his expeditions had captured. Beebe often shifts from analyzing a captured specimen to narrating the experience of being captivated by that very creature. This need not be seen as a mere ruse or distraction, as Koll suggests, but instead as a more capacious sense of what scientific inquiry should be, as well as an ethical regard for the specimen as a (once living) being in its own right. Nonetheless, Beebe's expeditions killed untold numbers of specimens—that is, animals rendered as objects for scientific scrutiny. In "The World of Bermuda Fish," he lists some of the team's methods, including shooting, trapping, harpooning, dynamiting, netting, and dredging. The expedition at Nonesuch, Bermuda, increased the "deep-sea fish catalogue" by "twenty-five hundred new entries, bringing the total to over twenty-four thousand."[66] Gruesomely, Beebe details how various fish suffer and die once captured. Yet it seems Beebe did not undertake this slaughter lightly, as he left behind many indications of his kind regard for nonhuman life. In "Mid-Ocean," for example, Beebe praises "a kind hearted entomologist" who followed behind a scientist who was upturning stones, in order to "carefully" replace "every upturned stone." And because Beebe was friends with Teddy Roosevelt—infamous for his big game and safari hunting—his comments on a particular deep-sea fish are notable: "Yet we can never visualize even these viper-fanged, ravening creatures as killing for lust of killing. . . . No deep-sea fish was ever photographed with its fin upon a fallen victim!"[67]

When Beebe attempts to reanimate some of his captures through his imaginative prose, he not only creates a sense of encounter for his readers but also extends a posthumanist or animal ethics that insists on the liveliness, perspectives, and being of the creatures. An aesthetic and emotional appreciation, rather than the cold eye of objectification, is vital for an ethical relation to other beings. Beebe begins *Half Mile Down* by contending that even as "investigators" are rightly "obsessed by problems," "we are unfaithful to our trust if we allow biology to become a colorless, aridly scientific discipline, devoid of living contact with the humanities. Our intellects will never be equal to exhausting biological reality."[68] The arts and humanities are lively and vibrant, infusing science with color, keeping it from becoming a dry, detached pursuit. This "Apologia," which opens *Half Mile Down,* underscores the intellectual impossibility of "exhausting biological reality." Although biological reality will always exceed scientific capture, aesthetic and emotional responses, the domain of the arts and humanities, can put scientific pursuits in perspective: "We should all be happier if we were less completely obsessed by problems and somewhat more accessible to the esthetic and emotional appeal of our materials, and it is doubtful whether, in the end, the growth of biological science would be appreciably retarded."[69] Beebe writes nearly three decades after Roosevelt's 1907 essay concluded the so-called nature fakers controversy, which pitted science against sentiment. It is interesting to note that seventy years later, some of the scientists involved in the Census of Marine Life would echo Beebe, noting the utter impossibility of their task, to identify and count every creature in the sea. Chapter 3 discusses the epistemology of the 2000–2010 Census of Marine Life, as well as its inclusion of photography, videos, and other art. From Beebe in the early twentieth century to the Census at the turn of the twenty-first, the lavish oceans swamp taxonomic enterprises, inspiring aesthetic, affective, and speculative responses.

Conventional models of scientific objectivity, of course, prohibit "soft" or "sentimental" orientations toward their objects of scrutiny. Ironically, William Firebrace's description of the shape of Beebe and Barton's bathysphere sounds like a cartoonish depiction of disembodied scientific objectivity: a "detached eyeball, freed from any accompanying body, sinking down into the ocean."[70] The detached eyeball travels without the baggage of the physical or psychological self. Loraine J. Daston and Peter Galison, in *Objectivity,* contend: "All epistemology begins in fear—fear that the world is too labyrinthine to be threaded by human reason; fear that the senses are too feeble and the intellect too frail; fear that memory fades."[71] They argue that objectivity not only fears the immensity of the world but the interiority of the knower—in short, "subjectivity, the core self."[72] Beebe rejects such dualisms, performing as someone encountering deep-sea species, acting as a scientist, but also experiencing and conveying the sensations and aesthetic effects of these encounters. Even as his expeditions killed mountains of creatures as scientific specimens, his writings animate the fish and foster a sympathetic regard to marine life.

Beebe's accounts of the ocean depths were penned during the early twentieth century, when the science of ecology was developing alongside the aesthetic conservation movement and the utilitarian Progressive conservation movement. Writing about the development of ecology in the early twentieth century, Sharon E. Kingsland notes that "ecology, like its sister discipline genetics, was part of an effort to control life and to apply rational methods to a complex set of problems generated by the American desire to migrate into and adapt to new landscapes. Ecology was in this respect a quintessential Progressive Era science emerging in response to rapid growth and transformation of the land. Ecology was a science called forth by change; it sought to render natural processes more predictable so that further change could be better controlled."[73] While Progressive-era science sought order and control, the scientific revolutions of the late nineteenth and early twentieth centuries—those of Sigmund Freud, Charles Darwin, and Albert Einstein—complicated such matters, as did the long tradition of natural history writing. Beebe contends that neither science nor philosophy is contaminated by contact with literature, even juvenile literature: "I hope there are some 'earnest seekers after the truth' besides myself who like occasionally to turn from Whitehead and Einstein, Eddington and D'Arcy Thomson to Kaa, Bagheera, Rikki-Tikki, the Walrus and the Gryphon, and back again without feeling they have degraded or scrambled their mental processes."[74] As editor of *The Book of Naturalists: An Anthology of the Best Natural History* (1944), which begins with cave paintings, includes Darwin and Thoreau, and concludes with Rachel Carson, Beebe endorses values that mix science and literature, skills and emotions, education and enthusiasm:

> To sum up, I present an ideal equipment for a naturalist writer of literary natural history: Supreme enthusiasm, tempered with infinite patience and a complete devotion to truth; the broadest possible education; keen eyes, ears, and nose; the finest instruments; opportunity for observation; thorough training in laboratory technique; comprehension of known facts and theories, and the habit of giving full credit for these in the proper place; awareness of what is not known; ability to put oneself in the subject's place; interpretation and integration of observations; a sense of humor; facility in writing; and an eternal sense of humbleness and wonder.[75]

Beebe included excerpts from Darwin's *Diary of the Voyage of the H.M.S.* Beagle in the anthology, introducing him by saying that he "seems somehow the most natural of naturalists, the most completely amateur, the greatest scientist who derived all his magnificent results from living animals, from observations and meditations on their lives, homes, habits—their natural history."[76] Amateur, as praise, implies that expertise can be stultifying, corralling the capaciousness of, say, from another context, the "beginner's mind."

In one selection, Darwin writes admiringly of Humboldt's "glorious descriptions" and their "rare union of poetry with science": "The delight one experiences in such times bewilders the mind; if the eye attempts to follow the flight of a gaudy butterfly, it is arrested by some strange tree or fruit; if watching an insect one forgets it in the stranger flower it is crawling over; if turning to admire the splendour of the scenery, the individual character of the foreground fixes the attention. The mind is a chaos of delight, out of which a world of future & more quiet pleasure will arise."[77] The scientific eye is arrested and distracted by an aesthetic encounter with a "gaudy butterfly" or "strange tree," yet this "chaos of delight" will settle into "quiet pleasure," and science will not be harmed. While Beebe no doubt was a popularizer, publicist, and fund raiser, it may still be the case that he countered the development of Progressive-era ecology, which placed narrow notions of reason in the service of control. Seeing Beebe solely in terms of how he failed to deliver scientific data on his bathysphere descents obscures other aims involving a more philosophical sense of epistemology, ethics, and aesthetics, as well as an expansive sense of what science could be—and indeed what science had already been. Beebe, who treasured Lewis Carroll's *Alice in Wonderland* as well as the works of Darwin and Whitehead, intermeshes science and story, the empirical and the fanciful. Sounding rather like Donna J. Haraway's feminist epistemological critiques of the transcendent, God's-eye view, as well as the psychonauts who reference *Alice in Wonderland* as a psychedelic and ecodelic[78] text, Beebe describes looking for "little creatures" in the jungle: "Instead of looking down from on high, being apart, with titanic brush of bristles ready to capture the fiercest of these jungle creatures, I, like Alice in Wonderland, felt myself growing smaller, becoming an onlooker perhaps hiding behind a tiny leaf or twig."[79] Shrinking oneself, as a human and as a scientist, enables enchanted worlds to come into view, a prelude to multispecies epistemologies and ethics.

Laura Dassow Walls, in her study of Alexander von Humboldt, *The Passage to Cosmos,* argues that recovering Humboldt "locates the first global wave of environmental studies just prior to the era of scientific specialization."[80] Walls underscores the value of art in Humboldt's theory: "where it matters most, science must necessarily fail . . . where science must weigh and measure, abstract and bring away, art can make present to the senses and the imagination the fundamental experience of contemplating nature in its wholeness, generating a similar emotional impact."[81] Beebe, following Humboldt, exclaims in *Beneath Tropic Seas* that even though we can "coldly analyze and apprehend" the sight of fish pouring out of the jelly, "we must stop to give an incredulous, unscientific gasp of wonder."[82] Such a gasp, a bodily response to the creature's beauty, intermeshes the viewer with the scene through the sudden intake of breath, a momentary punctuation in the temporal flow. This "unscientific gasp of wonder" evokes Isabelle Stengers's question regarding Whitehead's "passionate intensity' ": "How can one address experience in the sense in which it resonates in this exclamation, that is, before it is invaded by the words that judge and distribute what belongs to the subject and what belongs to the object?"[83] An aesthetic, intrasubjective experience happens before the sorting into categories that distance, divide, and corral. The gorgeous, haunting, and artful black-and-white photograph of the Quad Jelly-fish, with the twelve fish that had been living inside it now swimming around it, appears on the facing page of *Beneath Tropic Seas,* allowing readers to feel something of Beebe's experience. Beebe himself suggests the relays between literature, art, and the aesthetic sense of captivation when he states a few pages later that he regrets his "derogatory" comments about the "free swimming" and "stinging" sea worms: "We have a literary catch-phrase,—'he was transfigured with emotion'—which may be applied most literally to these neried worms of the sea."[84] Such transfigurations saturate encounters of sea life with the surprise of novelty, the sense of being awash in unknown worlds, and the pleasures of becoming with others, especially those whom one does not expect and cannot adequately understand.

Within the notes for an unfinished manuscript, "Mid-Ocean," Beebe begins with the thrill of discovery but concludes with an intuitive, holistic epiphany that can only be degraded by reason:

> Out of the jar of life from four hundred fathoms—a good half mile—comes a maelstrom of scarlet shrimps, a mass of white

> round-mouths, a giant salmon jelly and here and there an oval rainbow of a Sapphirina. A slender black and white creature slips out and settles in a free corner, and I realize that it is something which no man has seen before. I lift it with all gentleness into a glass dish and focus upon it on my microscope stage. First impressions are always worth-while,—they are all too soon sicklied over with the blinding glare of conscious intelligence; each fact becomes reasonably related to another and the spontaneous whole is gone forever.[85]

Touch affords an intimacy here as Beebe "lift[s] it with all gentleness." While the "microscope stage" should bring knowledge, instead, it seems to separate through its blinding focus. Steven Shaviro, discussing Whitehead, explains that "feeling, as such, is the primordial form of all relation and communication," adding that an "act of feeling is an encounter—a contingent event, an opening to the outside."[86] When Beebe subordinates the "blinding glare of conscious intelligence" to the ephemeral "spontaneous whole," readers may wonder what that understanding could have been before it vanished, but Beebe's frustration actually leaves readers with their own sense of wonder, aligning them with his experience rather than making them repositories of digested information. Shaviro argues that because feeling "changes whatever it encounters," it is "irreducible to cognition": "It isn't anything we already know."[87]

Yet the practice of science proper can be hindered by what Beebe calls emotional appreciation, as when he confesses his squeamishness in dissecting a specimen: "As usual I hesitated a long time before making the first incision in the skin. This is something childish or incurable which was born in me,—a fundamental reluctance to spoil anything perfect or unusual. However, I overcame my [Beebe left a big blank space here in the typing, then later wrote in the word "scrupulosity" by hand] and began slowly and carefully to cut through the tissue wall."[88] The blank space in his account parallels the pause in the procedure. The space subsequently filled by the uncommon word "scrupulosity" suggests Beebe's reaction exceeded the parameters of proper scientific practice. The word itself gestures toward a conflict between conscience and scientific procedure, even as respect for the bodily integrity of an animal is cast as mere childishness. Objective reason cannot simply squelch more diffuse emotional, aesthetic, and ethical responses. The living creature haunts the specimen, and in turn, the specimen incites animating speculations about this and other creatures.[89]

In his notes on *Lophodolus,* likely meant for the "Living Lamps" chapter of "Mid-Ocean," Beebe writes, "As various fish come up from the black depths they are greeted with explanations indicative of the effect their sudden appearance makes on the observer, some are awesome, others ugly, others beautiful, but Lophodolus is sheer comic—and comical in a wistful, infantile way."[90] Beebe continues this detailed and entertaining description, comparing the fish to a "caracature *[sic]* of a Martian," noting its absurd body and marveling at its "great beacon" of light. The poetic description of the encounter with the fish concludes with a speculative numerical array, listed in two columns: "We might diagram this fish's chances in life as follows: Sight 10, Feeling 5, Nostrils 10, Armature 5, Mouth and teeth 10, Color Protection 10, Size Protection 10, Speed 5, Scattered Light 0, Concentrated Light 35."[91] Personal reactions, poetic description, and numerical speculations ripple through each other, a spectrum of aesthetics and science, data and whimsy, fragile and fluid "circulation of reference."[92] Venturing a guess at the "fish's chances in life," Beebe inhabits a personal piscine drama, an evolutionary micronarrative without evidence or metrics from natural history or science. Two different DTR artists, Helen Damrosch Tee Van and Else Bostelmann, provide visual animations of *Lophodolus,* with cartoonish and anthropomorphic versions of rather dejected but adorable creatures occupying the bottom right of their group scenes.[93] Beebe's designation of *Lophodolus* has since been "corrected" to be less melodious. The World Register of Marine Species includes "*Lophodolus lyra* Beebe, 1932" (note the musical "lyra") as an "unaccepted name," listing "*Lophodolos acanthognathus* Regan, 1925" ("thorned jaw" rather than "lyra") as the accepted name.[94]

Mexican painter and caricaturist Miguel Covarrubias painted several images of Beebe, all of which suggest improper relations between Beebe and his specimens (Plate 1). In one image, published in *Vanity Fair* in 1933, Beebe sits in the lab, surrounded by specimens in jars, guiltily curving over a fish in a pan with his back to the microscope, salting the specimen he should be dissecting, his own bodily desires diverting him from proper scientific practice. The caption reads, "Professor Beebe, gourmet and ichthyologist, secretly fries his new discovery instead of pickling it for posterity." In his 1928 work "Scientist Dr. William Beebe," Covarrubias depicts Beebe underwater, with undulating limbs, a fishy face, and large finlike ears, his body gracefully flowing with the slope of the sand, the shape of the coral, and the fish that school around him, calmly becoming fish. In a third painting, published in *Vanity Fair* in April 1935, August Piccard's huge globular head floats in the sky while Beebe's piscine visage is submersed (Plate 2). Two mermaids swim to greet Beebe, a conventionally gendered vision of the sea that Beebe himself did not foster. Below him lurk sea creatures who look like they drifted over from an Else Bostelmann painting. (A citation of her distinctively surreal style that will be discussed below.) Covarrubias's caricatures enact a stylized undoing of the human as Beebe becomes one of the marine creatures he studies, suggesting a submersed scientific practice that swamps the subject/object divide, dramatizing Beebe's desire to merge with the fish.

Moreover, while the scientific specimen may seem to promise solidity and stasis, that is not necessarily the case. Beebe describes two "scarlet ctenophores"—gelatinous creatures, hauled from a depth of eight hundred fathoms—that cannot be preserved; only the aesthetic resonance, captured by Beebe in this handwritten account, remains: "Nothing will preserve it and another morning will see it dissolved into formless fluff, with the water stained deep vermillion."[95]

The Voice of a Deep-Sea Fish

Beebe's writing, like Humboldt's, often foregrounds the "voice and subjectivity of the scientist, always making his choices, actions and feelings present in the text as he tried to read order into the bewildering complexity of nature, tried to record, translate and mediate its many voices."[96] According to Dassow Walls, however, such a practice harmed Humboldt's legacy: "As science rigorously excluded the aesthetic and emotive, Humboldt dropped from the cutting edge to the cutting-room floor, edited out of the narrative."[97] Beebe, writing nearly a hundred years after Humboldt, performs not only his process of encountering astonishing species but also speculates about their perspectives.

A charming document in the Princeton University Archives playfully honors Beebe's devotion to the practice of imagining piscine perspectives. "The toast to William Beebe. Birthday Dinner July 31st 1931" begins, and takes as its chorus, the line "I am but a voice! The voice of a deep sea fish!"[98] Beebe's most notable fish tales are not narratives of murderous man making but instead portals that transport readers into animal worlds, scientific speculations akin to Jakob von Uexküll's "forays." In his *The Arcturus Adventure* (1926), based on his second expedition to the Galápagos Islands, Beebe writes: "My own interest in plankton is wholly that of trying to disentangle the lives of some of the small people—to put myself in their places by day and night, but I feel that I must establish their importance in the minds of more practical and far-seeing readers."[99] While science demands an assessment of practical importance, Beebe is driven by his curiosity about the lives and perspectives of plankton as "small people," speculating about their tangled relations. Defending his practice of merging the aesthetic and emotional with science proper, Beebe argues: "If I should consider the deep sea and its inhabitants solely from the point of view of technical science, I could never use such words as terrible, strange, beautiful or ugly. Because, in the essence of things, a pressure of a ton on each square inch is only a normal shift in physical conditions,—a fish which is chiefly mouth is merely a specialized adaptation to its particular environment, as is the smaller organ of our brook trout. But in this chapter we may let emotional appreciation go hand in hand with truth, and science will take no harm."[100] While he does not explicitly argue for an expansion of ethics to the nonhuman world, his invocation of "emotional appreciation" and his practices of speculating from nonhuman perspectives provide welcoming habitats for developing animal ethics. Beebe might well be echoing Whitehead's *Science and the Modern World,* one of the books he brought on his travels. Arguing that nineteenth-century literature "is a witness to the discord between the aesthetic intuitions of mankind and the mechanism of science," Whitehead asks, discussing Wordsworth, "Is it not possible that the standardized concepts of science are only valid within narrow limitations, perhaps too narrow for science itself?"[101]

Indeed, the route to imagining creaturely perspectives merges science, philosophy, and narrative and thus, not surprisingly, is neither narrow nor well trodden. In "Living Lamps," which was to have been a chapter in "Mid-Ocean," Beebe writes: "I have seen no medieval representation which is so extreme [illegible] as this deep-sea fish. And I am the only living human being who has seen one alive."[102] But seeing it is not enough; he wishes to see from its perspective. What he wouldn't have given "to have been told what these eyes had seen in the black depths, what that great mouth had engulfed, what enemies had been avoided, what part the mighty luminous head had played in [illegible] or in war, only the huge fangs could be luminous, why a great [illegible] tree of hundreds of medusa-tentacles was necessary—why? Why? Why?"[103] The dramatic repetition crashes into the obdurate unknown, dramatizing the desire for a scientific practice that would reveal the perspectives, the lives, and the being of deep-sea species. Shaviro's writing on Whitehead is helpful here, both in terms of Beebe's quizzical exclamations and in terms of Graham Burnett's lament (discussed in the introduction), regarding his own epistemological defeat in the pursuit of knowledge about whales. Shaviro writes that whereas an "act of feeling is an encounter," cognition may result in the assumption that "my not-knowing is only a contingency for myself, that ignorance is a particular state I am in; they imagine that the object I am seeking to know is in itself already perfectly determinate, if only I *could* come to know it."[104] More specifically, from the perspective of animal ethics and posthumanism, it is crucial to add that no living creature and no species should be objectified into some *thing* that is "perfectly determinate." The vitality, intelligence, agency, and emergent activity of living creatures and their interactions underscores being as becoming, encounter as event.[105] Neither the impossibility of the knowledge that Beebe and Burnett seem to be seeking from the species they pursue nor the value of empathetic or other entangling feelings, however, undermines the significance of scientific, artistic, philosophical, and other speculative practices that would imaginatively populate the depths with creatures worthy of interest, understanding, and ultimately, within a later era of collapsing ocean ecosystems, palpable concern.

More prosaically, the researcher in the archives also encounters uncertainty. An anonymous Post-It in the Beebe archive about the "Mid-Ocean" notes, "Assorted materials, 1930s," "Contents: Essay material, probably much unpublished, toward a book to have been called 'Mid-Ocean.' . . . W.B. had planned all during the Bathysphere years, to write this book. It was to have followed *Half Mile Down.*" The chapter titles for the planned book frame the animals in terms of beauty, wonder, drama, horror, myth, and poetic devices: "Crystal and Silver Zones," "Deep-Sea Dragons," "Monsters in Eternal Night" (with sections on

"Evolution" and "Alice in Wonderland"), "Feelers and Peerers," "Living Lamps," and "The Terrible Bottom Life."[106] The book was not finished, and some of the handwritten notes remain illegible (at least to me). The project hovers, still murky to this researcher, who could easily echo Beebe's frustrated cries of "Why, why, why?"

No matter. The breadcrumbs in and out of the archive allow us to trace how a particular fish, the silvery hatchetfish, circulates as specimen, photograph, painting, and whimsical narrative. A black-and-white photograph of a "Silvery Hatchet-fish (*Argyropelecus* and *Sternoptyx*)"[107] appears in *Half Mile Down,* with the caption stating that these fish "were seen twenty eight times on various dives in the bathysphere. . . . These fishes have many lights and their bodies are covered with silver tinsel."[108] Like many of Beebe's captions, this one highlights the creature's beauty, which the stark black-and-white specimen photo does not seem to capture. Else Bostelmann amplified the aesthetic appeal of the hatchetfish in one of her most stunning paintings, "Silver Hatchetfish Drifting through Abyssal Darkness," which will be discussed later. This painting was published in the New York Zoological Society's journal and in *National Geographic.* It also became Beebe's New Year's card. (Much later, the painting drifted onto the cover of this book.) In Beebe's unfinished manuscript, "Mid-Ocean," the hatchetfish inhabits a verbal rather than a visual medium, as it becomes the protagonist of a speculative narrative, "Argyropelecus." Beebe narrates the life of a fish, "young Argy," from the time when he swims with his mother to the time when he is caught by a "famous ichthyologist" and put in a glass jar. Beebe begins by asking "who can prove that the story of a tiny fish is less important than our own?" and concludes by asking "who can say that their little silver souls were not conscious of the peering eyes that gazed down on them in wonder and admiration?"[109] He speculates about Argy's dark, abyssal world, where it was "the same beautiful black day for aeons," asking us to consider "worlds in infinite space" that may be "too all inclusive for finite minds." The first paragraph of "Argyropelecus" reads like science fiction:

> It was the same beautiful black day that it had been for aeons and aeons and would be for aeons and aeons to come. The blackness was lit by a firmament of moving worlds, [whose] glowing spots

> of myriad-hued lights glided and flashed through the hydrosphere. Uncharted and trackless were the courses of these worlds.[110]

Beebe echoes the idea that the perpetual darkness of the depths renders it a realm without time. The discovery of the ocean's deep scattering layer by the U.S. navy in World War II, as well as the subsequent understanding of diurnal vertical migration undertaken by myriad ocean species, however, ironically suggests that the imagination of Argy's world as timeless and trackless is itself historically situated. As it turns out, some species of *Argyropelecus,* including the species once called "silvery" but is now designated as "Sladen's," participate in diurnal vertical migration. Subsequent scientific captures do not invalidate Beebe's attempts at imagining the perspectives of deep-sea creatures, of course, a practice that populates the empty seas with alluring life and suggests that their perspectives are worth imagining.

Mapping the Depths: A Vertiginous Aesthetics

Beebe, in an ecological fashion, enmeshes the imagination of multispecies perspectives with habitats, seeking to map the vertical distribution of various animals. From Jacob von Uexküll to Thomas Nagel and beyond, the project of imagining what it is like to be another species has been a formidable, perplexing pursuit, but for fish, and especially for deep-sea life, the project is made all the more difficult by the fact that human language is overwhelmingly terrestrial. Beebe explains:

> I succeeded in merging myself with the life of the fishes, aided by the lack of fear or even respect with which they greeted my entrance into their world. But when I began to think in words I found that just as I had to have a stream of atmosphere flowing down to me, bringing with it all the little motes and beams belonging wholly to the upper world, so when my mind began resolving what my senses sent to it, into outflowing words, these were ever burdened with dry-earthly similes and metaphors.[111]

Arid human language is unsuited to the task of describing piscine life. Translations of aquatic experience are desiccating; language is limited by its terrestrial terrains. The scale of the seas, their breadth, depth, and unfathomable volume, as well as the formidable task of mapping where particular species live, result in imaginative mapping as a vertiginous aesthetics.

When Beebe descended in his bathysphere with Barton, he recorded his observations on the changing light and the animals he glimpsed through the window. He struggled to portray the profound beauty and strangeness of the animals he saw. He labors, in his prose, to translate a visual aesthetic apprehension into written accounts that will be compelling for readers. In his typed notes entitled "Dive Number X: Spectrum," he details how colors disappear as he and Barton descend to 800 feet. At 50 feet red is "invisible," at 200 feet orange is gone, at 300 feet "yellow green" is "almost gone," at 400 feet violet "has completely eclipsed blue," at 500 feet "every color gone but violet," and by 800 feet there is "no color—not a particle of the faintest color distinguishable by either Barton or myself."[112] Beebe was mesmerized by the indescribable color of the water, which underscored the fact that the fish inhabited a world utterly different from his own. As he and Barton descend in their first bathysphere expedition to seven hundred feet, Beebe attempts to register the liquid hue: "I brought all my logic to bear, I put out of mind the excitement of our position in watery space and tried to think sanely of comparative color, and I failed utterly."[113] He switches on the searchlight, using it as a miniature sun to normalize his color perceptions, but to no avail: "the blueness of the blue, both outside and inside our sphere, seemed to pass materially through the eye into our very beings. This is all very unscientific; quite worthy of being jeered at by optician or physicist, but there it was."[114] Logic, sanity, and imperviousness, the bedrocks of scientific objectivity, fail him as the color of the water "materially" enters their "very beings." Unable to directly experience the deep waters without the protection of the bathysphere, they nonetheless experience a direct encounter with the color—the light—that penetrates them.

In his many ruminations on the color of the water in *Half Mile Down*, Beebe mixes scientific, philosophical, and aesthetic perspectives, engaging in what might now be called "elemental thinking,"[115] as pondering the color of the water dissolves categorizations. Beebe notes, "I think we both experienced a wholly new kind of mental reception of color impression. I felt I was dealing with something too different to be classified in usual terms."[116] While Beebe's vague observations on the aquatic color spectrum hardly seem noteworthy, Henry Fairfield Osborn claims that the "two most surprising phenomena [of the bathysphere dives] were, first, the abundance of life observed, and the clarity and certainty with which it could be seen and identified, and second, the blue brilliance of the watery light to the naked eye, long after every particle of color had been drained from the spectrum."[117] On their seventh dive, Beebe continues to study "the changing colors, both by direct observation and by means of the spectroscope," recording each color as it disappears at a particular depth. At eight hundred feet, for example, he sees "only the deepest, blackest-blue imaginable," adding that "this unearthly color brought excitement to our eyes and minds" on every dive.[118] He waxes poetic about the color of the water throughout *Half Mile Down,* marveling at "a solid, blue-black world, one which seemed born of a single vibration—blue, blue, forever and forever blue."[119] Ravished and overcome by the blue-black of the waters, Beebe revels in the powerful strangeness, the indescribable difference of this watery world. While this may seem more poetic than scientific, if we consider that ecology illuminates the relations between species and their environments, we could understand Beebe as an experimental subject here, recording the sensations and effects the light has on him as he descends. The fact that these sensations are not elaborate or codified or authoritative is exactly the point. He is within an entirely new realm for which he lacks evolved or constructed concepts and frameworks. To descend, not transcend, means to immerse oneself in aquatic zones that scramble and disorient.

By 1931 (if the handwritten notation on the archival document is correct), Beebe was able to construct a vertical chart divided into six depths, with "Pelagic Fishes" and "0–300 Fathoms," the deepest of which contained at least thirty-three species, which was included in the notes for "Mid-Ocean." While that chart was most likely based on specimens that they collected, the chart from the later bathysphere dives is more mysterious. Organized by depth, from shallow to deep, the typed notes entitled "Unknown Animals" seem to have been intended to vertically map bioluminescent creatures observed through the bathysphere window. The notes are organized by depth, starting at 260 feet and concluding at 3,028 feet, with each entry noting what was observed, followed by the dates, from 1930 to 1934. This archival document differs from the published appendixes in *Half Mile Down,* which includes more comprehensive notes from dive numbers 30 and 32. In a concluding section entitled "Unknown Animals" of the penultimate appendix of the book, Beebe suggests why he did not publish these notes on "Unknown Animals": "I have one hundred and fifty-four separate and distinct notes on Unknown Fish, and two hundred and thirty-five notes on Unknown Animals. To list these would most excellently reveal my abyssal ignorance of the majority of sparks, lights, half outlines, and glimpses of heads, tails, or eyes." He provides a brief excerpt of these notes, however, "which will be of value only when I or some other diver descends and resolves them into something understandable."[120]

His "abyssal ignorance" is productive, however, as the archival document on "unknown animals" possesses an aesthetic value, underscoring a brilliant, lively, and elusive aesthetic. At 570 feet, there are "no sparks yet," then at 590, "three little dim flashes," at 840 "first brilliant spark," at 875 "biggest and brightest flash ever seen," at 1170 "a network of light," at 1200 "millions of lights; many bright ones and many pale green ones," at 1670 "decidedly blue for the first time," at 2090 "now ghostly things in every direction" and "like meteors in every direction," at 2500 "lots of stuff," at 2800 "marvelous lights."[121] Along with these vertical notations of bioluminescence, Beebe charts mysterious creatures, relying heavily on similes, to visually translate the unknown life-forms into terrestrial human terms: "something like a rocket bursting," "something like a huge necklace of silvery lights," "something wriggling like mad," "again lace-like luminous creatures like a lot of wide-mesh nets." Beebe draws on outer space and desert landscapes to evoke the depths, comparing the underwater scene to "a whole desert zone of pitch blackness" and "desert darkness with black hole," casting the bioluminescence as "meteors" or "jet black comets." One observation, "lovely bright solid pale blue light close to glass. Probably ———," leaves a string of hyphens where the name of the creature could be—but wasn't—filled in later. The sparse notes are saturated with the delirious ecstasy of the unknown. Despite the fact that Beebe was separated from the life outside the bathysphere, the notes include a moment of intimacy as the creaturely light lands on Beebe: "Very brilliant flash; John saw this reflected on W.B.'s face."[122] Beebe's observations—so often dominated by the captivating colors of the seas and its creatures—embody a nascent posthumanist philosophy in which human knowledge systems and terrestrial horizons are overwhelmed. Dry scientific epistemologies of rational objectivity become drenched with embodied aesthetic and emotional responses. The gathered data drift, eluding systems or grids. Even if some of these writings were eventually rendered into properly scientific essays, the experience of reading them in the archive dramatizes the fresh and disorienting encounters with perplexing beings that elude taxonomic grids and volumetric mapping. Beebe's accounts of seeing unknown creatures from the bathysphere window are often mesmerizing, sensitive, and poetic: "suddenly a vision to which I can give no name, although I saw others subsequently. It was a network of luminosity, delicate, with large meshes, all aglow and in motion, waving slowly as it drifted."[123] Beebe performs a dance through knowing and bewilderment, mindful of his responsibility to convey something more scientific than the "Oh's! and Ah's" of his "first few dives,"[124] yet refusing to tame his vertiginous, aesthetic experience of encountering the creatures and lifeworlds of the depths. A half century later, such exclamations would be echoed—but tamed and contained—by those who descended in *Alvin,* as chapter 2 will discuss.

Perhaps Beebe's deep-sea scientific legacy has less to do with a conventional sense of scientific success in capturing specimens, collecting data, and making data yield illuminating structures and systems. Instead, he performed rather than represented, reflecting on the experience of being a terrestrial human who was overcome and undone by his immersion in the the depths. Beebe refused to divide the sensational and aesthetic from the scientific and objective, the arts and humanities from the sciences. The creatures may have eluded his scientific capture, but they nonetheless sparked more intimate interrelations, thus making his failures ethically and epistemologically illuminating, as they reflected new worlds as well as more capacious modes of being and knowing that exceeded the confines of rationality.

The persistent aesthetic dimension of Beebe's encounters with denizens of the seas may be understood through Romantic traditions. Forest Pyle defines "radical aestheticism" from the Romantic poets and beyond in *Art's Undoing: In the Wake of a Radical Aestheticism:* "A radical aestheticism returns us to the aporias between perception and sensation, on the one hand, and cognition and conceptualization, on the other; and it forces us to behold the insuperable nature of these aporias, the aporias out of which the aesthetic as a 'domain' is constituted in the first place."[125] Perhaps the accelerating divide between science and the arts plunged Beebe into such aporias, unbridgeable gaps between "perception and sensation . . . and cognition and conceptualization." Beebe, who included many scientific illustrators in the DTR team, seemed to reject a representational model that would constitute the aesthetic as a separate domain, fostering fluidity between these realms. Melanie Sehgal underscores that the aesthetic, for Whitehead, is not restricted to "specific kinds of experiences," but instead, "the aesthetic is a *specific dimension* inherent to all kinds of experience, again taken in a non-humanist sense . . . that takes into account and puts into practice and motion the reciprocity of feeling, relationality and existence."[126] The "reciprocity of feeling, relationality and existence" may be a prelude to a disanthropocentric, environmentalist, multispecies ethics that can arise from aesthetic experiences. Sehgal concludes her analysis of Whitehead, Stengers, and Haraway thus: "Philosophy in the sf mode as aesthetic fabulation, works in a performative rather than representational way, inventing lures for feeling in view of different kinds of becoming. Aesthetic concerns are crucial for such an endeavor: changing thought requires and presupposes changes in habits of feeling."[127] What sorts of changes would these be, in Beebe's case, given the scale of violent specimen collecting and the lack of environmental concern for ocean ecologies at that time? Beebe's scientific practice entailed harming countless animals, yet if his writing about marine life shifted habits of thought and feeling, the direction would be toward curiosity, regard, and even concern for the creatures as beings with their own perspectives. Beebe casts deep-sea creatures as aesthetically potent beings, ultimately unknowable yet worthy of speculative attention. The gasps of wonder, delight, astonishment, and captivation may entice readers to feel the confines of science proper and to aesthetically appreciate and speculate about each of the creatures in the sea as well as the lifeworlds they inhabit.

Interlude: Deep-Sea Slither and Gender Drift

One review of *Half Mile Down* identifies the "weakness" of the book as arising from trying to "please two distinct groups of readers, the lay public and the scientists."[128] Interestingly, in the preceding review, published on the same page of the 1934 volume of the *Nation,* Margaret Mead recommends Ruth Benedict's *Patterns of Culture,* promising that it will "delight alike the social scientist, the student of art, and the sensitive-minded lay reader," without trepidation about the mixed audience.[129] While Beebe contended that science can withstand emotion, aesthetic responses, and popularization, as we saw above, his position was entangled in gendered dualisms. Marcel C. LaFollette, in *Making Science Our Own: Public Images of Science, 1910–1955,* argues that most depictions of scientists in magazines during this period "implied that success in scientific research required certain 'masculine' attributes such as intellectual objectivity, physical strength, and emotional detachment."[130] Even poetry, presumably the "softer" opposite of science, was charged by Ezra Pound, along with other masculinist modernist writers, with the task of becoming "harder and saner": "The poetry which I expect to see written during the next decade or so, it will, I think, move against poppy-cock, it will be harder and saner, it will be what Mr. Hewlett calls 'nearer the bone.' It will be as much like granite as it can be, its force will lie in its truth, its interpretative power (of course, poetic force does always rest there); . . . At least for myself, I want it so, austere, direct, free from emotional slither."[131] Pound longs for a poetry that escapes the taint of the frivolous feminine and the degraded "slither" of creaturely emotions,[132] whereas Beebe embraced both the "emotional" and the "slither," dodging gendered dualisms and disciplinary divides.

At a time when scientific research was obviously dominated by men, Beebe hired and supported the careers of many (presumably white) women. Mark Dion, Katherine McLeod, and Madeline Thompson write that not only did the popular appeal of the DTR lead to "skepticism in academic circles," but so did "the fact that the DTR staff included women as part of their professional research team."[133] McLeod notes that Beebe "facilitated Rachel Carson's first helmet dives in the Florida Keys in 1949."[134] Beebe "encouraged the careers of influential scientists such as Sylvia Earle and Rachel Carson."[135] And he hired many female artists, including "Isabel Cooper, Anna Taylor, Rachel Hartley, Helen Damrosch Tee-Van," and the extraordinary Else Bostelmann.[136] Beebe's employment of women "was unusual and drew criticism."[137] McLeod notes that men called his "inclusion of women in these spaces a de-professionalization of the field."[138]

Gloria Hollister, who was Beebe's chief assistant, not only held the "women's record" for her 410-foot descent in the bathysphere (a surprise for her birthday) but also invented a method for using ultraviolet rays to make fish specimens transparent. The *Science News-Letter* devotes six paragraphs to Hollister's innovation, only to taint it with this caption underneath a photo of a transparent fish specimen: "Ultraviolet, used by bathing beauties to give them a becoming coat of summer tan, has been used to the opposite effect on this fish, in a process which makes him transparent for study."[139] The scientist's achievement is trivialized by the objectification of feminine beauty, as well as the passive voice that erases Hollister's innovation and the gendering of the fish as transparently, "universally" masculine. The biographical sketch of Hollister included in the Beebe archives notes that Hollister's father, who was a physician, encouraged her love of nature and did dissections with her but wouldn't allow her to become a doctor because it was unfitting for a woman; yet he encouraged her to become a scientist. In later years, Hollister was instrumental as a conservationist, protecting the Mianus River Gorge in New York, the beginning of the Nature Conservancy. Hollister's notebooks in the Wildlife Conservation Society's Department of Tropical Research Collection include titillating newspaper clippings about her talks, with headlines such as "Girl Finds Thrill in Exploration."[140]

Beebe begins his strange essay, "The Jellyfish and Equal Suffrage," by admitting it is "a long cry from the jelly-fish to equal suffrage," and yet it is also "a long cry from the moon to the tides." Echoing Darwinian feminists of the late nineteenth century Antoinette Brown Blackwell and Eliza Burt Gamble, who diminished the significance of sexual difference in order to argue for women's rights,[141] Beebe looks to the marvelous sex of the jellyfish, arguing that "in this land of uncertainties, sex does not lend itself to an earnest, philosophical consideration." He continues: "It exists indeed, but when any given individual may be of either sex, or none whatever, it is difficult to take the question seriously."[142] Beebe defends gender equality through a kind of "gender minimizing feminism,"[143] in which sexual difference is as diaphanous and insubstantial as gelatinous creatures, asking readers to think with these aquatic species in a manner that is fantastically inclusive and expansive. Although his serpentine argument, which moves from the jellyfish to consider other creatures, is confusing (it both endorses and critiques sex specialization in animals, including humans), it echoes feminisms that pose nature as an undomesticated ground,[144] releasing us from the oppressive categorizations of culture. While "nature has shown herself just and generous," men and women have been "victims of a perhaps too complex civilization."[145] In the final paragraph, he presents this fanciful proclamation of genderqueer beings as a political force—"The voice of the jellyfish is heard throughout the land demanding equality in all things"—only to conclude with a more domesticating heteronormative vision of the "wild gander and his capable and cooperative mate, the goose."[146] But readers may rejoice in the fact that the jellyfish has already left the barn, venturing on to more queer and feminist seas.

Else Bostelmann: Surreal Seas

Else Bostelmann (1882–1961), Beebe's most distinctive expedition artist, fashioned an enduring surreal aesthetic for deep-sea life. Exposing the lie of transparent, mimetic objectivity, her work captures the specimen in detail while captivating viewers by enticing them to speculate about life in the depths. Whether or not she knew of the work of French biologist and surrealist Jean Painlevé, whose short films about sea life, starting in the early twentieth century, used scale, sound, and narration to enhance the strangeness of shallow-water marine life, Bostelmann's distinctive paintings, many of which were published in *National Geographic,* were aligned with Beebe's sense of abyssal life as enticing and ineffable, as well as with other surrealist encounters with the depths. According to Madeleine Thompson, director of the library and archives at the Wildlife Conservation Society in the Bronx Zoo, where most of Bostelmann's paintings reside,[147] Painlevé gave Beebe a seahorse photo, and Salvador Dalí visited the zoo a few times.[148] These personal connections pulsate with broader currents in surrealism, an avant-garde movement that aligned art and science, and that echoed earlier natural history traditions that Beebe himself embraced. Donna Roberts explains: "Like other Romantic naturalists, such as Goethe and Alexander von Humboldt, the study of nature required an active and imaginative subject, one that responded to feeling as much as reasoning. It is this combination of the Romantic naturalist that is reconstituted within the figure of the surrealist as emotive and analytical, ecologically engaged and self reflexive."[149] While Beebe himself did not project a surreal sensibility, he epitomizes the Romantic naturalist, and he treasured Bostelmann's surrealist deep-sea illustrations.

Mary Ann Caws states that surrealism "aimed above all to preserve a sense of the extraordinary, the unexplained and the inexplicable." As an "art based on desire," it "requires a necessary otherness, or it withers into boredom."[150] That encounter with otherness allows surrealism, broadly conceived, to be understood as an apt style for an inhumanism, posthumanism, or xenophilia. A surrealist aesthetic, consistent with the deep descents that Beebe undertook, does not rise above and attain a comprehensive vision of an expansive landscape. Instead, it descends into a world of immediacy, unreason, and creaturely encounters. Maurice Nadeau, referencing Einstein but concluding with Freud, writes:

> The epistemologists fall into step, questioning the conditions and limits of knowledge. . . . Reality is something besides what we see, hear, touch, smell, taste. There exist unknown forces that control us, but upon which we may hope to act. We have only to find out what they are. Man torn between his reason—discredited but still arrogant—and an unknown realm which he feels to be the true source of his acts, his thoughts, his life, and which has been revealed to him in the sleep that devours nearly half his existence, man dares turn his eyes upon it. He becomes acquainted with strange creatures; he moves in landscapes never seen before; he performs enthralling actions. A Viennese psychiatrist, armed with a dark lantern seeks to penetrate the dim labyrinth.[151]

While these landscapes and creatures are in this case merely metaphorical because they symbolize the Freudian unconscious, they could be cast instead as metonymic, evading the domestication of the nature/culture divide, heading off to more murky realms. Painlevé's surrealist and scientific films of sea creatures, starting in 1928, pose aquatic life itself as subversively strange. James Leo Cahill, in *Zoological Surrealism: The Nonhuman Cinema of Jean Painlevé,* argues that Painlevé's films contributed to "unthinking anthropocentrism," arguing that surrealism "drew upon but also cultivated a disposition of radical openness to the strange, unexpected, and new."[152] The octopus slithers and slides not only through Painlevé's films but also through other surrealist works. For example, Guillaume Apollinaire's short poem, "Ocean of Earth" (1918), begins, "I built a house in the middle of the ocean / Its windows are rivers which flow out of my eyes / Octopus stir all around its walls."[153] The ocean, with its unknown depths that readily symbolize the unconscious and its diversity of disorienting creatures, seems a rich realm for surreal dreams. Dalí, for example, donned a diving suit at the International Exhibition of Surrealism in 1936[154] and created a titillating surreal aquarium funhouse, the Dream of Venus, at the 1939 World's Fair, where Bostelmann's paintings of deep-sea life "were on prominent display."[155] Dalí's funhouse, filled with bare-breasted women lying about in beds or swimming as mermaids, was tiresomely conventional in terms of its depiction of women as spectacles.[156] Less predictable is Jindrich Styrsky's 1934 "Cuttlefish Man," a pink biomorphic shape with a vaginal opening and a long serpentine appendage on a watery blue background, which suggests the possibilities for post humanist or inhumanist surrealisms in which the human psyche is not confined within the individual; instead, the human itself is improperly constituted with a disconcerting array of parts that merge human with nonhuman, ungendering the being in the process.[157]

Bostelmann was not alone in depicting ocean life as surreal, but her signature aesthetic for deep-sea life in particular—strange, spirited creatures against a black background—popularized the sense of deep-sea life as eerie and enticing, forging an enduring surreal aesthetic of the depths.[158] Little is known of Bostelmann, who trained as an artist in Germany and exhibited her work in several German cities before immigrating to the United States, where she married and gave up painting, studied science, and later became an illustrator for Beebe and other scientists. *National Geographic* published "more than 300 plates of her deep sea and shore fish and 104 water-colors of sea life," not including her watercolors of whales.[159] Her artistic methods are notable: she consulted specimens and photos and verbal descriptions and even conducted in situ observations while painting with oils underwater at fourteen feet, wearing a diving helmet.[160] As Natascha Adamowsky notes, Bostelmann "was given a host of heterogenous medial sources and practices—text, spoken language, sketches, photographs, and specimens—which she then had to synthesize visually."[161] Bostelmann describes her swift capture of the colors of the rapidly changing specimens: "I undertook the quick sketching of their fleeting, iridescent hues, their still glowing light organs, the ephemeral brilliancy of their metallic sheen."[162] Although the Bermuda National Gallery featured an exhibit focusing on Bostelmann in 1996, *Bostelmann Paints for Beebe,*[163] it is the twenty-first century that has brought a resurgence of interest in Bostelmann's art, perhaps sparked by a broader interest in deep-sea life but certainly propelled by historian Katherine McLeod's extraordinary array of articles and interviews and the 2017 exhibit at the Drawing Center in New York City, entitled *Exploratory Works: Drawings from the Department of Tropical Research Field Expeditions.* The exhibition was curated by McLeod along with Dr. Madeline Thompson, the Wildlife Conservation Society's director of library and archives, and Mark Dion, known for his animal art. Bostelmann's work has not only been promoted within artistic and cultural venues but also, in an odd turn of events, in scientific circles. Marine biologist Edith Widder participated in the Lighthouse Art Center's 2016–17 exhibition, *Illuminating the Deep,* in Tequesta, Florida, which featured paintings by Bostelmann held in a private collection. In an article, "The Fine Art of Exploration," published in *Oceanography,* Widder tells how she was having breakfast with a friend, Janeen Mason, who said she had "a bunch of old paintings" in her closet, including some "fish paintings by her husband's great grandmother."[164] It turns out these were by Bostelmann, whose work Widder first encountered in *Half Mile Down.* Widder is certainly not alone in her regret that Bostelmann's journals "were destroyed in a flood."[165] After examining many of Bostelmann's works in person, Widder praises her meticulousness, especially in terms of color, explaining how Beebe, despite his verbal skills, was challenged by the impossibility of describing "the utter strangeness of all he saw" through the bathysphere window.[166] Widder concludes with her wish that Bostelmann's art will be exhibited in more accessible venues, such as the "Ocean Portal at the Smithsonian's Natural History Museum, or better yet"—she wildly dreams—"how about our very own Smithsonian Museum of Oceanography?"[167]

The *Exploratory Works* exhibition catalog notes that the Department of Tropical Research produced more than two thousand illustrations, and that within the DTR, "artists were not simply decorators of the scientist's writings; instead, they were essential communicators who understood their role in the dissemination of information about ecological relationships."[168] This is no doubt broadly true; however, some of Bostelmann's depictions of deep-sea life do much more than convey information. They are brilliantly composed to create a sense of encounter between the viewer and the deep-sea creatures, some of whom seem to be swimming up to see the viewer. The aesthetic is eerie, enticing, potent, and sometimes playful, luring us into contemplating the being and lifeworlds of these extraordinary creatures. Bostelmann's work takes a distinct stylistic turn when she shifts to black backgrounds for the deep-sea environs. Beebe's essay in *National Geographic* in January 1932, "The Depths of the Sea: Strange Life Forms a Mile below the Surface," details the trawling methods used to capture the specimens, along with several black-and-white specimen photos taken by Beebe, many of which have black backgrounds, and seven paintings by Bostelmann of groups of fish, often with their prey. These seven paintings, which include species caught at the surface, in midwaters, and in the depths, are all on white backgrounds. While Bostelmann depicts the specimens as lively beings before capture—they chase their prey, bodies curving in movement—their watery milieu, which would have been different shades of blue and black, depending on the depth, is absent until the final painting. The eighth painting, "The Tangled Web of Death of the Deep-Sea Squid," features a fish tangled in the grip of one of two squid, with long, coiling tentacles, against a solid black background, with a caption that notes the "abyssal darkness."[169] Granted, it would have been difficult to distinguish abyssal black fish, such as "The Dragon of the Shining Green Bow," with its extraordinarily long gray chin barbel, against a black background, yet in some paintings, Bostelmann outlines the black fish with glints of light, making them less visible as separate specimens and more metonymic with their abyssal worlds.[170] The black background that Bostelmann began to use for deep-sea life is distinctive as it reverberates with a surreal aesthetic that suggests the deep seas cannot be readily illuminated.[171]

Illumination is an apt metaphor for understanding deep-sea life, especially in terms of depicting Beebe and Barton's bathysphere descents into the dark waters, where they peer out of small windows. Two rather different paintings of the bathysphere demonstrate the potential for the abyssal realms to decenter anthropocentrism. Bostelmann's untitled painting depicting the bathysphere on a shallow contour dive, just forty feet down, features a school of "many thousands of Giant

Blue Parrotfish" swimming past three windows of the bathysphere, where they would have been readily seen by Beebe and Barton.[172] The peaceful light blue fish swimming along in bright aquamarine waters nearly cover the bathysphere. Given that the fish swim between the humans in the bathysphere and the human viewers of the painting, they are positioned within a tame terrain of human understanding in bright surface waters. However, in "Bathysphaera Intacta Circling the Bathysphere," the bathysphere is dwarfed by two enormous "untouchable bathysphere fish" that swim around it (Figure 2).[173] The bathysphere hovers in a realm where human knowledge is limited; the light beaming out of the window fails to illuminate the bulk of the dark water surrounding the submersible. One human face is barely visible in the middle portal. Beebe and Barton have a restricted view, while the fish, with their carefully painted, shining eyes and brilliant white teeth, are a strong, formidable presence. The eye of the fish near the bottom is painted blue, red, pink, black, and white, a microcosmic reflection of its vibrant bioluminescent environment. Scientific observation cannot capture the fish themselves; the vastness of the seas provides ample room for them to see the vessel without being seen. Even as the viewer is afforded a perfect perspective on the creatures, the scene erodes epistemological foundations, as the viewer's perspective only seems omniscient. The speculative scenario allows us to see what Beebe and Barton cannot quite discern, yet their experience of being submerged in the depths, as well as the vaster waters of the fish themselves, cannot be adequately conveyed. Anthropocentric, terrestrial omniscience and objective perspectives are surreal, a visual fiction that obscures the obscurity of the depths. While the deep-sea creatures can be made to signify things, even things such as the unknowable, they can also speak to an irreducible, nonsignifying potential that brings the depths into the domain of human concern while acknowledging that such worlds are untranslatable, impervious to illumination. Bostelmann's surrealistic style perfectly embodies the oscillation between the need to portray specimens in a scientific manner, "adding reality, not subtracting reality," to riff on Latour,[174] and the sense that the creatures elude human understanding. Moreover, Bostelmann's distinctive style itself emphasizes a kind of making, a poesis, rather than a mirrorlike objective mimesis, that plunges scientists and artists into a realm of encounter, relationality, and endless speculation.

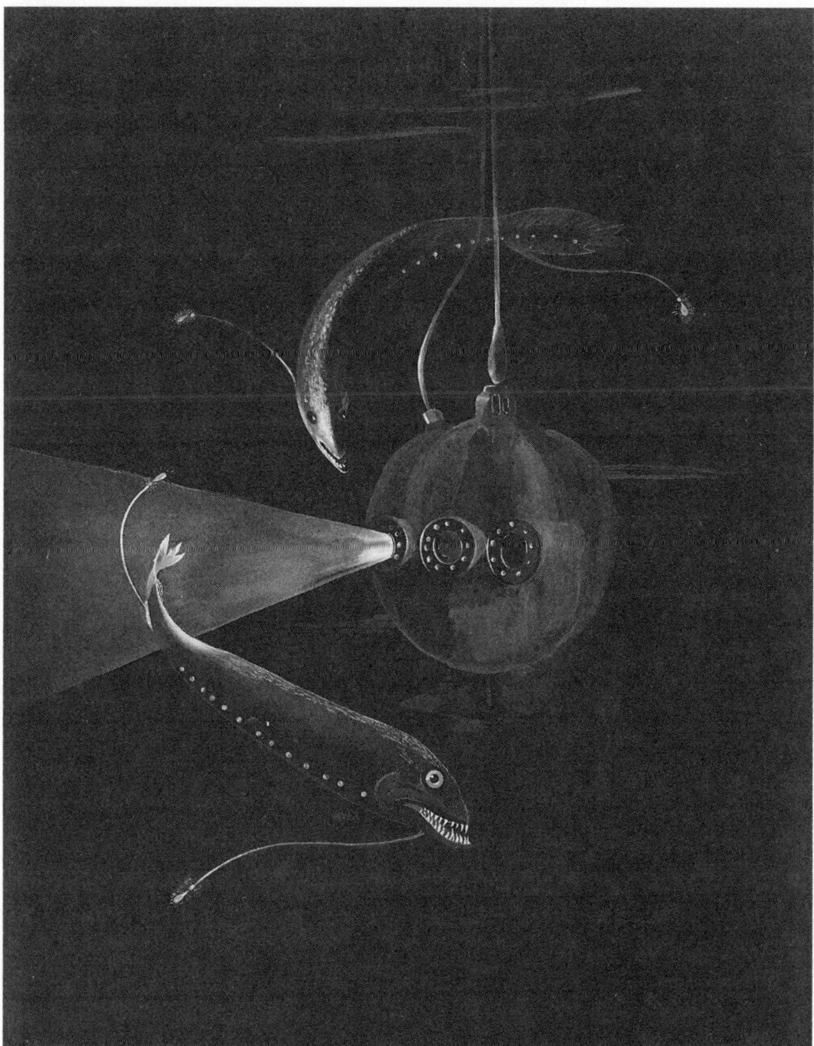

Figure 2. Beebe, from the portal of the bathysphere, glimpses just a bit of the tail of a giant dragonfish circling the submersible. Barely visible, long, eellike creatures blend into the waters above. Else Bostelmann, "Previously Unknown Giant Dragonfish *(Bathysphaera intacta)* Circling the Bathysphere," watercolor on paper, 1934, WCS Archives, WCS-1039-01-05-1002-R-CR. Copyright Wildlife Conservation Society. Reproduced by permission of the WCS Archives.

Bostelmann depicts a disanthropocentric surrealism in the sense of a heightened realism, which evokes the strangeness of existing entities—not in order to reveal the subconscious or unconscious human mind but instead to portray stunning abyssal creatures and speculate about perception and being beyond the human. Although this is a tall order, it is in keeping with Beebe's philosophical musings about the life of the fish he captured or glimpsed as well as his scientific interest in the eyes of the deep-sea fish (Figure 3). *"Stylophthalmus,"* for example, depicts three silvery fish, two with eyes attached to extremely long, thin poles, and one with giant eyes that seem barely attached to its head. Their morphologies seem whimsical and outlandish, an impossibility, with their eyes so precariously perched.[175] Bostelmann epitomizes what Arnaud Gerspacher calls the "artist as ethologist." He asks: "What happens to ethology and the study of animal behavior when the contours of its creaturely subject-object of study can only be traced, alluded to, formulated, touched, and communicated by an artistic jump or, perhaps, even a leap of faith? If this is the case—that certain ontological and epistemological truths can only be ciphered through judicious aesthetic fictions—then is this not what art practices can bring to ethology?"[176] Bostelmann, who took ten years off from artistic practice to study natural science before joining the DTR team, performs such breathtaking aesthetic leaps.

One of Bostelmann's most dramatic paintings zooms in on a sabertooth viperfish, sixteen times its natural size, its mouth wide open, the triangle of the jaws dominating nearly the entire frame, its huge fangs about to catch cheerful young sunfish (Plate 3). The giant eyes of the tiny sunfish, a mix of blue green, black, white, and yellow, with that characteristic spark of life, as well as the big eye of the viperfish, painted in blue, green, purple, and orange, with pink dots against black under the eye and a quite noticeable reflection of white light in the black iris, suggest their spirit, their sentience, their vivacity.[177] Captioned "A Fish-Eye View of a Microscopic Tragedy" in its *National Geographic* printing, the painting shrinks the viewer, placing us, along with the adorable sunfish, in the jaws of the viperfish. Our gaze is directed to the empty black space directly in the middle of the scene: the spot where the jaws will shut. The flat black background conceals the waters beyond, leaving no escape. A dual aesthetic is in play: the tiny fish, with their giant eyes that humans can't help but see as cute, contrast with

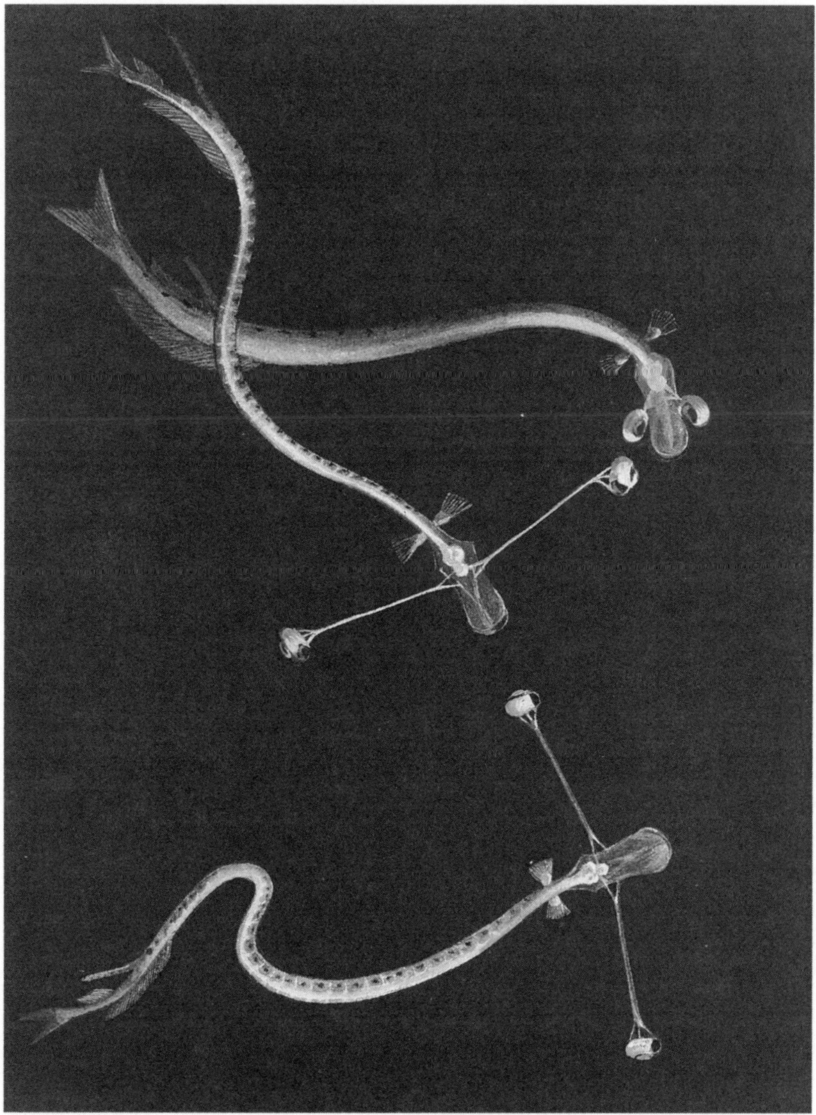

Figure 3. One of the surreal fish paintings by Bostelmann that dramatizes the strange eyes—and by extension the inconceivable perceptions—of the fish. The elegantly curving bodies depict lively movement and perhaps companionship, while the two huge eyes that are not on extensions seem to peer at the viewer. Else Bostelmann, *"Stylophthalmus,"* watercolor on paper, 1930–32, WCS Archives, WCS-1039-01-05-U081-R-CR. Copyright Wildlife Conservation Society. Reproduced by permission of the WCS Archives.

the surreal, frightening dragonfish, captioned as an "ogre," with its impossibly wide mouth and long, pointed fangs. Ecological understanding precludes rooting for the prey over the predator, of course, but the overstated "microscopic tragedy," as well as the expressions of the fish, infuse the scene with whimsy and ironic drama. The huge teeth, filling half the scene, are more graceful than menacing. Bostelmann's artful painting, captioned "Orange-Lighted Finger-Squid Catching Lanternfish," taken from "a full mile depth," features "a pair of enormous eyes," one of which is on view.[178] The slender, spotted squid gracefully arcs on the diagonal as the tentacles ripple, curl, and catch the lanternfish. The giant eye of the squid, its diameter nearly the width of the squid's body, placed just off center in the composition, looks directly at the viewer in an unnerving gaze. Another painting that features eyes, perhaps Bostelmann's most utterly gorgeous work, is an untitled painting of two green fish with bright red stripes and radiating silvery fins. Widder included a full-page copy of this painting in her article on Bostelmann, identifying the fish as *Dolichopterix,* or barreleyes. The geometric eyes—blue barrels lodged in a transparent head—could not be more surreal. Swimming on the diagonal activates the pair, while the way their heads come together expresses an intimacy between the fish that spills over to the viewer.[179]

While studying Bostelmann's paintings and drawings in the Wildlife Conservation Society archive, I was struck by their exquisite detail and color as well as by the devoted attention she paid to the eyes, made up of multiple colors but always containing a spark of light, reflecting the light they were seeing as well as suggesting their own perceptions of their own worlds. Such attention to the eyes populates the abyss with lively, sentient, perceiving creatures. Bostelmann's renditions are consonant with Beebe's emphasis on the eyes as symbolic of sentient being. In the "Fish Outline" for the Bothidae, he writes: "The visible focus of its whole being is a silver gilt iris, circling a pupil as black as its haunts." Imagining the being of the specimen, he is pained by its capture: "With difficulty we realize that two hours ago this paper thin being was swimming amid the joy of icy blackness."[180]

In one of Bostelmann's most magnificent and most surreal paintings, "Silver Hatchetfish Drifting through Abyssal Darkness," a group of hatchetfish confront the viewer with their striking eyes (Figure 4).[181] The abyss stares back! Three of the hatchetfish appear brightly silver and yellow; three more lurk in the background, dark bluish gray, nearly blending into the black ocean. The diagonal arrangement of the fish is dynamic; while five of the fish swim to face the viewer directly, one heads up and away. Beebe must have had a particular affection for this painting. Not only was it included as a full-page, full-color illustration in "A Half Mile Dive in the Bathysphere" printed in the *Bulletin of the New York Zoological Society,* but as mentioned previously, he also used it as a New Year's card.[182] Exposing the lie of transparent, mimetic objectivity, Bostelmann's work captures the specimen in detail while captivating us with its stylized strangeness. The almost geometric, modernist eyes—the eye at the top looking exactly like a piece of metal hardware (a cap nut, to be precise)—tempt us to muse about what it is these fish see. That commonplace piece of hardware, astonishingly out of place, punctuates the surreal aesthetic, scrambling domains by exporting something ordinary from land to a distant and radically different world. And yet the caption lures us to connect with these fish as beings, or even, anthropomorphically, as a kind of people, as it states that their eyes are not only "enormous" and weirdly "pointing upward," but also that "in front they have a singularly human aspect."[183] Their dramatically downturned mouths seem, if human, rather dejectedly so, but in an ominous and edgy, not doleful, manner.

Carol Grant Gould, in her biography of William Beebe, writes that Bostelmann's paintings of deep-sea life from the bathysphere dives "revealed creatures so unearthly that many critics refused to believe in their existence."[184] After her work appeared in *National Geographic,* "Beebe was labeled an amateur by some hardheaded scientists, and a romantic and a fantasist by many others."[185] Gould notes, however, that "in years to come almost every creature Bostelmann painted from Beebe's descriptions would be seen and identified."[186] Bostelmann's striking art had the paradoxical effect of both threatening Beebe's scientific reputation and gaining attention for the bathysphere dives. In the glaring light of land, surrealism may seem a mere fantasy, a separate realm from objective scientific knowledge, part of a fairy land of eerie beauty that is bewildering, discomfiting, even alarming. But if such creatures are actually real, what then? Mark Fisher argues that the weird "involves a sensation of *wrongness.*" He explains: "a weird entity or object is so strange that it makes us feel that it should not exist, or at least it should not exist here. Yet if the entity or object is here, then

Figure 4. With eyes like pieces of hardware, five of these surreal fish with gloomy expressions seem to skeptically scrutinize the viewer. Three of them can barely be discerned as they blend into the dark background, creating horizontal depth. Else Bostelmann, "*Argyropelecus hemigymnus:* Silver Hatchetfish Drifting through Abyssal Darkness," WCS Archives, WCS-1039-01-05-U077-R-CR, 1929–31. Copyright Wildlife Conservation Society. Reproduced by permission of the WCS Archives.

the categories which we have up until now used to make sense of the world cannot be valid. The weird thing is not wrong, after all: it is our conceptions that must be inadequate."[187] But when abyssal species that are hauled to the surface become stories, photographs, and paintings, what could, what does, "here" even mean? Are the deep seas an alien realm, cordoned off somehow from the rest of the planet? Where is the "here" of these renditions of deep-sea life? John Wyndham's novel *The Kraken Wakes,* which directly references Beebe's bathysphere descents, and which depicts an epistemological maelstrom provoked by possibly alien abyssal creatures, addresses some of these questions, as I describe next in chapter 2. Chapters 3 and 4 will also discuss the vexing trope of deep-sea life as alien, which confuses the depths with outer space and casts abyssal life outside the terrain of environmental concern.

While some of the sea life Bostelmann captured was too "unearthly" to be believed, other depictions could be seen as lacking in proper objectivity because of their anthropomorphism. The surreal morphs into the cartoonish in "Deep Sea Angler Fish," which depicts two dramatically large rows of white, seemingly smiling teeth (not unlike those of a hastily made jack-o'-lantern), opalesque eyes, and the lure, which looks like a nose, against a black background that subsumes the fish's body. The cartoonish image slips away from being a scientific rendition, delighting in whimsy.[188] Yet the playfulness in the cartoonish images may serve an ecologically edifying purpose. Beebe, in a typed archival essay entitled "Abyssal Life," writes that he is often asked whether, given the extreme cold and pressure of the depths, "all play or natural exuberance is absent in such abnormal conditions." He responds that "the answer hinges upon the interpretation of the word 'abnormal,'" noting that the environments of abyssal fish are, of course, normal to them. He adds, "When I see a delicate jellyfish pulsating happily along at 2000 feet, I am sure it possesses all the capacity for joy (if any) and life that endows a corresponding jelly at the surface."[189] Depicting the jellyfish as potentially playful and joyful in this context stresses the fundamental interrelations of species and environment as well as the species' own distinctive being and vitality. The same image of the deep-sea angler fish reappears along with seven adorable pals in "Front View of 8 Deep Sea Fish," also captioned "Big Bad Wolves of an Abyssal Chamber of Horrors" (1934), a title that playfully casts the specimens as fairy-tale figures and dispels dread, disgust, or thalassophobia.[190] Who could fear this playful, jovial bunch (Plate 4)? With their eyes directed straight at the viewer, the group of cheery fish seem to have swum up to greet us. The abyss, a surprisingly jubilant locale, looks back. Bostelmann's strategic anthropomorphism, with its staging of encounters, fosters an ethical regard for the abyssal animals. This shift from the formidably surreal to the more comfortably cartoonish in Bostelmann's work could parallel Painlevé's shift from an objectively scientific perspective to that of an appreciation gained by committing something like a crime. Cahill explains that in the 1920s and 1930s, Painlevé, with his scientific background, was wary of anthropomorphism, often apologizing for the anthropomorphism in his films, but six decades later, Painlevé made "a rather astonishing proclamation: 'We commit anthropomorphism. We have the right to commit anthropomorphism. We have the duty to commit anthropomorphism. Otherwise we would be incapable of appreciating any element around us.'"[191] The fact that anthropomorphism has long been an offense to science proper reveals the human exceptionalism and anthropocentrism upholding conventional ideals of modern Western scientific objectivity. Primatologist Frans de Waal alerts us to the opposite problem, that of "anthropodenial," or the refusal to recognize any "human" traits in other species—traits, talents, abilities, and characteristics that are shared across human and nonhuman animals.[192] It is telling that the measure of truth and the criteria for reality reference the dead specimen and the stiff fact while aesthetics, emotion, liveliness, and a sense of encounter and relation are rendered suspect. There is something fishy going on when vivacity is restricted to human being. Ethical regard for other beings and passion for protecting environments are fostered by encounters with other species. Deep-sea life, which ordinary people can only encounter in profoundly mediated ways, depends on aesthetic depictions to provoke interest, care, connection, and concern. If those depictions are surreal—adding to, rather than subtracting from, the real—and deeply disorienting and unmooring, they are all the better for developing disanthropocentric imaginaries and abyssal posthumanist visions.

Bostelmann developed an aesthetics of the abyss that paralleled other surrealist projects and manifested Beebe's philosophical stance on the relation between science and the arts. The distinctive abyssal aesthetic she created was provocative, effectively condensing many of the philosophical and scientific quandaries pertaining to life in the depths in ways that continue to capture the popular imagination. In terms of contemporary scholarship on posthumanism, critical animal studies, and the blue humanities, her work remains good to think with. Cary Wolfe in *What Is Posthumanism?* provides a succinct frame for understanding the ethical orientations of Beebe and Bostelmann's transdisciplinary work in terms of how they welcomed the new beings of the abyss: "[It] will take all hands on deck, I think, to fully comprehend what amounts to a new reality: that the human occupies a new place in the universe, a universe now populated by what I am prepared to call nonhuman subjects. And this is why, to me, posthumanism means not the triumphal surpassing or unmasking of something but an increase in the vigilance, responsibility, and humility that accompany living in a world so newly, and differently, inhabited."[193] While Wolfe was not referencing Beebe, Bostelmann, surrealism, or deep-sea species, and while the DTR expeditions killed a horrific number of nonhuman subjects in the pursuit of science, the work of Beebe and Bostelmann nonetheless manifests what may now be understood in terms of a posthumanism that provokes epistemological humility in the face of worlds "so newly, and differently, inhabited."

After working with Beebe, Bostelmann collaborated with A. Remington Kellogg, a cetacean scientist who had never himself actually seen the whales he wanted Bostelmann to paint, to produce a "full-color gallery of every known cetacean" published in *National Geographic* in January 1940.[194] Collaborating with Kellogg, Bostelmann painted "a gallery of colorful, vital, and even *playful* whales and dolphins, nursing their young, gamboling sociably among their kin, and frolicking in grotto like shadows."[195] Burnett describes the public's response as "overwhelming." The success of the article "reveals the degree to which Remington Kellogg succeeded in shaping how Americans came to 'see' whales and dolphins at midcentury," not as monsters but as beautiful, playful, social creatures.[196]

While Bostelmann's two distinctive aesthetics of deep-sea life may seem contradictory—the eerie, disanthropocentric, unsettling surreal and the cartoonishly anthropomorphic—they merge into a place where the human is not the only sort of being or person, and the world is not devoid of surprises. Both aesthetics can foster a sense of connection to the creatures; both attest to the necessity for scientific and aesthetic captures as well as the murky border between them. Writing about her practice of painting from specimens, Bostelmann states: "Sometimes I let them face me—and I am astonished at how often they resemble one or another of my friends! Lots of fun they are." This essay, originally published in *Country Life* magazine, concludes with a domesticating transfer of the deep-sea creatures into mere terrestrial decoration. Even though the "mysterious realms of the deep ocean" have remained "unknown to us," we can now, she states, bring such "artistic beauty . . . within reach of our modern life" and "enjoy it as part of our daily existence" by using it to decorate "our summer homes."[197] The unsettling sense of unfathomable depths, disanthropocentric surrealism, and a metonymic slide between strange creatures and their worlds is diluted by this cozy, privileged, merely decorative aesthetic. The conception of the Anthropocene, in which nearly every species, habitat, and planetary system has been disturbed or destroyed by colonialism, industrialization, capitalism, and more, makes the idea of abyssal species being hauled up from their worlds only to become common ornamentation even more frightful. Bostelmann's illustrations, haunted by the creatures they depict, however, continue to circulate, in exhibitions and digitally. They entice viewers who, in the early twenty-first century, inhabit a world in which even the deep seas have been irrevocably harmed, a world where nearly everything is flattened into an inert resource for use. Yet deep-sea life, as it circulates through the writings of William Beebe and the art of Else Bostelmann, retains its ability to enchant and astonish. In 2022, nearly a century after the bathysphere descents, Ray Troll and the Ratfish Wranglers, in "The Ballad of William Beebe," sing of the surreal and psychedelic life of the sea, with eerie background vocals and the refrain "Ocean creatures that would blow your mind, blow your mind!"[198] Reveling in surreal and mind-expanding aesthetic encounters is infinitely preferable to the boredom, militaristic manufactured ignorance, and clumsy captures that I discuss next in chapter 2.

2

Packed Up in Tupperware and Russian Vodka

Deep-Sea Science Fiction and Nonfiction

> To get the feeling of what it is like to be a creature of the sea requires the active exercise of the imagination and the temporary abandonment of many human concepts and human yardsticks. . . . We cannot get the full flavor of marine life—cannot project ourselves vicariously into it—unless we make these adjustments in our thinking.
>
> —Rachel L. Carson, *Under the Sea Wind*

> There is nothing obviously glorious about a tubeworm on deck. Pulled out of its tube and drained of its color, the worm is decidedly flaccid and unattractive. The vibrancy of the live worm on the seafloor is lost; in the atmosphere of the sea surface the plume loses its vitality and pales.
>
> —Cindy Lee Van Dover, *Deep-Ocean Journeys*

The *New York Times* reported in January 1944, "The famous bathysphere used by Dr. William Beebe for the study of deep sea marine life has gone into war service."[1] World War II ended Beebe's bathysphere descents, concluding this remarkable period of deep-sea exploration that merged science, art, narrative, and philosophical rumination, accentuating the aesthetic dimensions of marine species and the vertiginous process of imagining abyssal worlds. While Beebe's accounts of ocean exploration influenced the science writing of Rachel L. Carson as well as the science fiction of John Wyndham and others, deep-sea exploration in the United States during World War II, the Cold War, and beyond became a technological, utilitarian, and political matter, driven by military interests and the practical concerns of the navy, especially in terms of its submarines. Expect no surreal paintings, whimsical stories, or speculations about creaturely lives and perspectives to dance through this chapter. Robert Ballard, a marine geologist who spent thirty years at Woods Hole Oceanographic Institute after serving in the navy, writes that "no one really continued in Beebe's footsteps as he had hoped." Instead, "a long inactive period, from 1934 to 1948," followed the bathysphere dives, perhaps because, he suggests, "most countries concluded that deep-sea exploration would never yield commercial benefits."[2] Military exploits, capitalism, and extractive industries rule; natural history traditions are left behind. Marine biologist Tony Koslow describes the next decade, however, as a crucial time for deep-sea discoveries: "The exploration of mid-ocean depths began in earnest in the 1950s, as a by-product of the Cold War development of submarine-based military technology. The desire to conceal its submarines at sea and to find the enemy's turned the U.S. Office of Naval Research into a major patron of American oceanography; the deep scattering layer, bioluminescence, and marine acoustics were all deemed to be of military as well as purely scientific interest and readily attracted research funding."[3] Naomi Oreskes, in *Science on a Mission: How Military Funding Shaped What We Do and Don't Know about the Ocean,* is less sanguine than Koslow, however, demonstrating that "purely scientific interest" was not prioritized. In fact, she demonstrates how military funding resulted in the production of ignorance about marine biology. Oreskes includes one particularly telling document, a memo from the Scripps Institute of Oceanography from 1945 with a chart listing nine research topics, including "currents," "tides and tidal currents," and "bottom sediments," along with their military value, such as "mine warfare," "fleet operations," and "salvage." The last item on the list is "biology and chemistry," cramming massive fields of science into a single category, with the purpose of understanding "fouling and corrosion," presumably of naval vessels and accompanying hardware.[4] Oreskes argues that while the usual story of the astonishing discovery of hydrothermal vents was a "triumph of curiosity-driven research," the navy constructed *Alvin,* the submersible that delivered these discoveries, not for "basic research but to satisfy the military demand for deep submergence capability."[5] Indeed, when a marine biologist requested to use *Alvin* to study the giant squid, he was denied. This lack of interest in marine biology by those who held the purse strings had rippling effects. After the war, the biologists at Scripps, according to Oreskes, lobbied for a biologist as the next director, to "redress the imbalance created by wartime concerns" and "restore equity for biological work."[6] Shockingly, revisions to the graduate curricula at Scripps "eliminat[ed] marine biology altogether," focusing instead on "basic problems of the sea." Oreskes asks, "Was life itself not a basic problem of the sea?" and "Was it not one of the *most* basic problems of science?"[7] Similarly, by the 1950s at Woods Hole, "biology was being relegated to second-class status."[8] Oreskes is careful to qualify her argument, noting that some research on fish was done at other places with other funding, and some biological research was conducted at oceanographic institutions—but even that was driven by military concerns, such as enlisting dolphins for military operations. Given the dire state of ocean ecologies in the twenty-first century, Oreskes's compelling contention that the lack of basic understanding of marine life was actually produced by economic, political, and military forces is painfully on target.

> With the U.S. dominating global scientific oceanography, and the needs of the U.S. Navy dominating U.S. oceanography, oceanographers in the late twentieth century constructed reliable knowledge about the ocean as a physical medium through which sound and submarines might travel, but they also constructed substantial ignorance about the ocean as an abode of life. There is no reason in principle why attention to physical oceanography should have crowded out good biological work, but in practice, it did.[9]

It was not until the Pew Charitable Trusts and the Alfred P. Sloan Foundation launched the ten-year Census of Marine Life, spanning 2000 to 2010, the topic of chapter 3 in this book, that marine life finally received considerable attention. But as Oreskes puts it, the Census "came too late": "Scientists were trying to count something that to a large extent was already gone."[10] We can extend Beebe's speculations about the lives and environments of deep-sea creatures to imagine a surreal bestiary of all the deep-sea species that must have existed in the early twentieth century but were rendered extinct as a result of extractive industries, pollution, and other anthropogenic causes by the end of the twentieth century. The fact that deep-sea creatures have eluded capture for most of human history naturalizes the manufactured ignorance of the midcentury. But now, when deep-sea technologies enable the "discovery" of new species, manufactured ignorance of anthropogenic marine extinctions remains the norm, as assaults on marine ecologies accelerate, largely unseen.

In its analysis of science writing and science fiction penned by authors from England, Canada, and the United States in the long period between Beebe's bathysphere descents and the Census of Marine Life, this chapter identifies two opposing currents: the aesthetic marveling provoked by deep-sea life, which is oriented toward environmental visions, and the containment of aesthetic responses in the service of objectification and extractivism. It is worth repeating Melanie Sehgal's interpretation of Whitehead: "Rather than referring to specific kinds of experiences, the aesthetic is a *specific dimension* inherent to all kinds of experience, again taken in a non-humanist sense that the philosophy of Whitehead or also James suggests, that takes into account and puts into practice and motion the reciprocity of feeling, relationality and existence."[11] While the "reciprocity of feeling, relationality and existence" orients us toward posthumanist, environmentalist, and multispecies ethics, the containment of the aesthetic as a contaminant of the scientific process strengthens reason as mastery, dominant dualisms, and human exceptionalism. While many scientists, science fiction writers, and nonfiction writers (sometimes the same people) experience deep-sea environments and abyssal species as mind-blowingly aesthetic events, the ideologies of big science, skewed toward military domination and extractive capitalism, repel the "reciprocity of feeling, relationality and existence." The brutal reduction of living creatures to specimens, as well as, ironically, the failure to capture deep-sea specimens, contains and discards aesthetic experiences. As the aesthetic flows through science, science writing, and science fiction, it frames deep-sea life as intrinsically valuable. Then, through unthinkably immense metonymic speculations, it disperses this regard though the volumetric seas.

A Placid Prelude: Nonhuman Explorers in Rachel Carson's Tranquil Abyss

I would be remiss if I began this chapter with anyone other than Rachel L. Carson, who published three magnificent books about the ocean in the 1940s and 1950s: *Under the Sea Wind* (1941), *The Sea Around Us* (1951), and *The Edge of the Sea* (1955). While Carson did not write extensively about the deep seas, the third section of *Under the Sea Wind,* "River and Sea," which Beebe included as the final selection in *The Book of Naturalists,* follows eels from the river to a "strange world," "a deep abyss near the Sargasso Sea, where they spawn and die."[12] Carson intended the book to make "the sea and its life" a "vivid reality . . . out of the deep conviction that the life of the sea is worth knowing."[13] While Carson's prose is treasured as natural history and nature writing, her speculations about the being of other species, not unlike those of Beebe discussed in chapter 1, cannot be corralled into science proper, as they venture into speculative fiction, science fiction, or speculative animal studies. Most importantly, however, given Beebe's protestations in the previous chapter, along with the work of Humboldt, Darwin, and Whitehead, readers may approach Carson's multispecies speculations without dividing science from narrative or knowledge from aesthetic experience. Similar to Beebe's critiques of the constrictions of modern science, Joni Seager sees Carson's emphasis on wonder as part of her "deeply intentional challenge to normative science": "She counterposed the clinical investigation of the laws of nature against an openness to revel in the beauty and marvel of the world."[14]

The sense that science would not be harmed by beauty and wonder echoes chapter 1, but here, we will consider how Carson presents the perspectives of marine species, connects the deep seas with waters more familiar to terrestrial beings, and shifts scales both temporally and geographically to defamiliarize anthropocentric visions. Carson departs from the supposed objectivity of modern science by recognizing that other species are not merely objects for human use and understanding but instead subjects with their own knowledges, viewpoints, and stories. Her discussion of the importance and the difficulty of imagining other species' modes of being precedes contemporary debates in critical animal studies. In *Under the Sea Wind,* for example,

Carson explains the dynamic tension between bracketing anthropocentric assumptions so as to understand a creature on its own terms and a strategic anthropomorphism: "To get the feeling of what it is like to be a creature of the sea requires the active exercise of the imagination and the temporary abandonment of many human concepts and human yardsticks. . . . We cannot get the full flavor of marine life—cannot project ourselves vicariously into it—unless we make these adjustments in our thinking."[15] Yet "we must not depart too far from analogy with human conduct if a fish, shrimp, comb jelly, or bird is to seem real to us—as real a living creature as he actually is."[16] Paradoxically, Carson must insist that these creatures are in fact real, even though they appear within narratives that include anthropomorphic analogies—comparisons concocted in order to make them seem real to readers awash in human exceptionalism and the scientific objectification of nonhuman life. She must accomplish this, of course, without seeming feminine or sentimental.

The first two chapters of the third section of *Under the Sea Wind* craft a lushly detailed narration about Anguilla, an eel who lives in Bittern Pond, swims through a stream into a river, then into the sea, along the way encountering fellow eels, trout, octopus, an angler fish, shorebirds, and more. While the habitats and behavior of each of the species receive precise description, as the silver eels swim to sea, "they passed from human sight and almost from human knowledge."[17] Anticipating the refrain of twenty-first-century ocean conservation, Carson demonstrates that while human knowledge about deep-sea life may be meager, human destruction of the oceans is not. Ominously, "mile by mile," the fishermen "moved farther out, and finally their nets came up filled with food fishes. The wintering grounds of the shore fish—the summer fish of the bays and river estuaries had been discovered."[18] "Discovery" here is not a matter of aesthetics, taxonomy, or appreciation but merely extraction for use. "The trawls went down through the hundred fathoms of water; down from ice and sleet and heaving sea and screaming wind to a place of warmth and quiet, where fish herds browse in the blue twilight, on the edge of the deep sea."[19] Carson gently imports a menacing "heaving" and "screaming" from above, casting the disruption as elemental, only indirectly indicting the disruptive trawls and the fishers. The idyllic aesthetic of the scene, the browsing fish in the "blue twilight," entices readers to sympathize with the fish, peacefully dwelling in their own world.

While the eels' tale has so far been ornate with ecological particulars while spare in style, once they enter "the deepest abyss of the Atlantic," Carson reiterates the lack of human knowledge of the deep seas: "The record of the eels' journey to their spawning place is hidden in the deep sea. No one can trace the path of the eels. . . . Nor is there a clearer record of the journey of those other eel hordes. . . . No one knows how the eels traveled."[20] To the young eels, the abyss is also an unknown "strange world," with "strangely shaped" creatures, fishes that were the "strangest of all," and creatures that "crawled precariously over the deep and yielding oozes of the floor of the abyss."[21] Carson's rigorous attention to nonhuman life and ecology decenter the human, presenting a disanthropocentric narrative of abyssal "exploration" as seen through the eels' encounters with new life-forms. The aestheticized descriptions of deep-sea life, as well as the inclusion of particular species, echo Beebe's poetic accounts and Bostelmann's stylized illustrations: "Before the eyes of the eels, clouds of copepods vibrated in their ceaseless dance of life, their crystal bodies catching the light like dust motes when the blue gleam came down from above." The shrimps "often expelled jets of luminous fluid that turned to a fiery cloud to blind and confuse their enemies." Small fishes "clothed in a leathery skin" somehow shine "with turquoise and amethyst lights," gleaming "like quicksilver." A riot of diminutive beauties that scintillate even when effervescently vanishing, the "mirrorlike flanks of the hatchetfish reflected the blueness of the water and their outlines were lost in a shimmer of light."[22] By focalizing the aesthetic recognition of the marine life through the eyes of the eel, Carson centers the eels' modes of perception and aesthetic appreciation, encouraging us to speculate about the eels' experience and to extend this liveliness across all the species within the scene, from copepods to jellyfish to the unnamed fishes, each of which sees from its own perspective. Compressing her jeweled descriptions of an abundance of species within the space of a single paragraph, Carson intimates the expanse of possible ecological relations between them. The aesthetic here is not only concerned with each individual creature as a beautiful specimen or treasure for a cabinet of curiosities, but also dances across the entire scene—as it is seen—by "billions of pairs of black, pinprick eyes peering into the strange sea world that overlays the abyss."[23] The abundance of life is not still, is not merely a canvas for human contemplation, but is a zone of diverse beings that see each other and experience the abyss in ways we cannot grasp.

In a quintessentially ecological move, Carson concludes the eels' tale with their return to the rivers and ponds; the eels thus connect familiar bodies of freshwater to the mysterious depths of the sea, but without domesticating the deep. Rather ominously, the eels wait, "creatures of the deep sea, ready to invade the land."[24] The invasion is impressive in its scope: "Some would go on for hundreds of miles—creatures of the deep sea spreading over all the land where the sea itself had lain many times before."[25] An admirer of Beebe, Carson nonetheless reverses the narrative of exploration, as the deep-sea creatures comes ashore, a scenario not unlike *The Kraken Wakes,* which will be discussed in the next section, and not unlike countless horror films in which beasts escape their proper place. Here, though, all is as it should be (even if it is unnerving to humans) as the eels do what they do, within complex interconnected ecologies. As the deep-sea eels return, they inhabit their own maps, their own modes of charting their course, from abyssal realms back to the pond. Then, shifting from creaturely time to geological time, Carson leaves us with one final speculative scenario, returning the earth itself to the sea. As mountain ranges erode, "once more all the coast would be water again, and the places of its cities and towns would belong to the sea." Rather than domesticating the deep, bringing it within the solid terrain of human knowledge, the abyss reclaims and defamiliarizes the very ground of the human.

In "The Sunless Sea," a chapter from *The Sea Around Us* (1951), Carson describes the depths in familiar terms, which resonate with Beebe's descriptions: "deep, dark waters with all their mysteries," "endless night, as old as the sea itself," "the mysteriousness, the eeriness, the ancient unchangingness," a region that has "withheld its secrets more obstinately than any other," a place where the "unrelieved darkness has produced weird and incredible modifications of the abyssal fauna."[26] Despite her precise account of the science of abyssal life, Carson subordinates human knowledge to that of marine life, once again insisting on other-than-human modes of knowing and being. After noting that "William Beebe's impressions from the bathysphere were of a life far more abundant and varied than he had prepared to find," the next paragraph elevates the abyssal explorations of marine mammals: "The existence of abundant deep-sea fauna was discovered, probably millions of years ago, by certain whales and also, it now appears, by seals."[27] In her characteristically calm tone, Carson depicts the ocean as a realm of multispecies exploration, perspectives, and knowledges, thereby placing the exquisite content of the chapter within a posthumanist, multispecies framework of evolutionary time in which what seems new to today's human scientists was already discovered, millions of years ago, by various cetaceans and pinnipeds. Echoing the aesthetic of the abyss and its creatures as eerie and mysterious, Carson, without fanfare, presents an astounding subject for speculation—not only that of abyssal life itself, but abyssal life as known by other species, through time. Such knowledges reframe human comprehension as just one way of knowing, just one mediated perspective, within a kaleidoscope of knowledges. Inviting us to speculate with eels, cetaceans, and other marine life, Carson reiterates the mysteriousness of the depths for humans, but she scrambles the terrain of knowledge through multispecies perspectives and evolutionary temporal scales, creating a posthuman aesthetics of the abyss that is quietly, contemplatively surreal.

The Bathysphere Sleeps. The Kraken Wakes. Speculations Spiral.

John Wyndham's novel *The Kraken Wakes* joins Carson in her calm contemplations of other beings, but in this case, the creatures happen to be not just alien but perhaps actual aliens. Wyndham, a British author perhaps best known for *The Day of the Triffids* (1951), published *The Kraken Wakes,* a rather "cozy"[28] Cold War horror tale about threatening and intelligent deep-sea life, in 1953. (The U.S. version was entitled *Out of the Deeps.*) The novel not only explicitly references Beebe and Barton's bathysphere descents but also punctuates musings on the ineffability of abyssal creatures with the desire to capture specimens. While Beebe and Bostelmann sought to make deep-sea life visible, so that the narratives, photos, and paintings spark speculation about submerged creaturely worlds, Wyndham's creatures become the site for speculation because they elude visualization as well as the most elementary sort of taxonomic grids (Figure 5). This provided either a formidable challenge or an unlimited opportunity for the illustrators to create cover art; different editions depict wildly dissimilar illustrations of the creatures! However, musings about the ineffable creatures underscore the persistent current in the aesthetic reception of deep-sea life, that of the je ne sais quoi, as Margaret Cohen has argued.[29] Here, however, it is not only that visual captures of deep-sea life are encountered as being too marvelous, surreal, or utterly novel to be captured in mundane terrestrial language but also that despite hundreds of pages of novelistic discourse, readers, along with the characters, cannot concoct a visual image or a coherent sense of what these creatures are. The divergent cover art seems randomly pasted onto different editions of the novel, supplements that betray the lack of an image within. The novel's creatures epitomize Jacques Derrida's *différance* as their limitless signification spins out across the narrative by differing and deferring, differing and deferring, without being defined, harnessed, or contained by an origin, a stable reference point, a specimen, an image, or even a morphology.[30] Francesco Vitale argues in *Biodeconstruction: Jacques Derrida and the Life Sciences* that "*différance* is a geneticostructural condition of the life of the living and its evolution."[31] Similarly, Stefan Helmreich and Sophia Roosth's conclusion to their keyword essay on "Life Forms" notes that "the claim of isomorphism across diverse biological entities points to the hope that it may be possible to craft an encompassing theory of the biological, one that may, however, be ever out of reach, ever multiplying, and ever escaping."[32] The "whatsits" in the novel, which may or may not be biological, escape enclosure within what Derrida argues is the "asininity" of the absurdly immense category of "the animal." They also gesture toward an abyssal alterity that exceeds the abyss itself, as both a conception and a place. Dramatically, the whatsits leave the depths and come ashore.[33] The event of "xenobaths" erupting onto the human stage destabilizes the foundations of human knowledge while repopulating the planet with other intelligences, "alien" in one of Stefan Helmreich's elegant definitions of the term, "life forms whose place in our forms of life is yet to be determined."[34] Deep-sea species, which continue to epitomize undetermined forms of life, may orient us toward a je ne sais quoi consonant with an affirmative sense of biological deconstruction—an aesthetic recognition of the limits of human knowledge, which opens up limitless space for the knowing and being of a multitude of creatures that cannot be represented. The impossibility of representation

Figure 5. Various book covers of *The Kraken Wakes* (also titled *Out of the Deeps*), showing diverging depictions of the bafflingly amorphous aliens.

does not foreclose speculation. As Cary Wolfe contends, with regard to Wallace Stevens's poetry, "because there is nothing to represent, there is *plenty* to do—an endless amount."[35]

The Kraken Wakes begins with a short metachapter entitled "Rationale," set after the events themselves, in which a journalist proposes to "write an account of all this," and his wife, Phyllis, also a journalist, asks: "Without references or records?" He answers, "If anyone ever *does* read it, then he'll be able to have the pleasure of digging up the documentation—what's left of it. My idea is simply to give an account of how the whole thing appeared to me—to us."[36] Their ensuing debate about where to begin the story turns on which events the unnamed creatures had been proven to have caused. The rationale foregrounds questions of evidence, documentation, narrative point of view, and the murky waters where witnessing phenomena is a matter for science, personal narrative, and popular opinion. Tennyson's 1830 sonnet, "The Kraken," serves as an epigraph after the rationale. This curious poem never describes the kraken in visual or morphological terms. It sleeps an "ancient dreamless, uninvaded sleep," "far beneath in the abysmal sea." The poem references the idea that life in the depths eludes time and history, but it concludes with the kraken suddenly—and inexplicably, unless one reads it as a Judgment Day scenario, as critics have done—awakening, roaring, and dying "on the surface."[37]

The first sentence following the poem asserts, in a glib, singsong rhythm that undermines its credibility, "I'm a reliable witness, you're a reliable witness, practically all God's children are reliable witnesses in their own estimation."[38] In "Phase One," Phyllis and Mike, on a honeymoon cruise, spot five glowing clouds in the sky with "no real shape visible." As they investigate what could have caused this mysterious occurrence, the ship captain hesitates to register an opinion for the record, stating that "it doesn't do a seaman any good to get a reputation for seeing things."[39] Beebe would wryly agree. Mike ruminates about matters of proof and witness, musing that there must be "a great many people who go around just longing to be baffled, and who, moreover, feel a kind of immediate kin to anyone else who admits bafflement along roughly similar lines. . . . There are strata of bafflement."[40] The term "strata" suggests the vertical layers of ocean zones; paradoxically, however, these scientific mappings, with all their apparent thereness and geologic solidity, become a matter of pleasurable confusion.

This intellectual curiosity is strangely absent, however, when Mike and Phyllis, as journalists, witness the descent of a "bathyscope" five and a half miles into the depths of the sea. They reference Beebe's historic bathysphere descent, but they remain unimpressed by the event at hand—"a pretty boring affair," they sigh, sneaking off to smoke cigarettes on deck.[41] Watching from shipboard cameras, they witness a narrated scene reminiscent of Beebe's account of his bathysphere descent: "All black and dead now. Not a thing to be seen. Funny thing the way these levels are quite separate. Ah, now we can begin to see something below. . . . Squids again. . . . Luminous fish. . . . Small shoal, there, see? . . . There's—Gosh!—"[42] The unflappable journalists finally seem interested in the "nightmare fishy horror" that "gaped" at them "from the screen."[43] This was a prelude, however, to the real drama, when a "fish—or maybe something else kind of turtle-shaped" but certainly a "monstrous-sized brute"—appears. The voice goes dead. The screen goes blank. When the bathyscope is pulled up, they discover the explorers are dead and the metal of the cables has fused.[44] Sending down the unmanned "telebath" reveals a "large, uncertain, oval shape at the extreme of visibility," which was "always tantalizingly ill defined, and never quite well enough illuminated for one to be quite certain even of its shape."[45] The allure of the amorphous propels the narrative, words chasing down but never catching what cannot be categorized or visualized.

Oddly, after the screen goes blank, the narrator asks, taking refuge in the coziness of patriarchal coupledom, "Why not praise one's wife?" explaining that "Phyllis can write a thundering good script."[46] The attention to mediation here, from the cameras on the bathyscope and telebath to the transformations from experience to script to novel, are a fitting postscript to Beebe's deep-sea excursions in that questions of evidence, witnessing, circulation, and spectacle remain at the fore. Moreover, while the journalistic duo hardly shares Beebe's sense of wonder for marine life, the sophisticates' tendency toward blasé boredom is punished by the plot: the deep-sea creatures—wreaking massive havoc on humans—are not to be ignored.

There are worse things than British boredom. The Americans launch a nuclear bomb five miles down, near the Mariana Islands. As the journalists witness this event, they see a cloud that "writhed and convolved upon itself in a fashion that was somehow obscene as it climbed monstrously up the sky."[47] The cloud embodies both a suffering creature and the transcendent agent of that suffering: the iconic nuclear mushroom cloud. The personification is apt because in the next paragraph, the question of whether an "intelligent" creature has been sinking the ships is raised. Also, the revenge-of-nature motif looms, as one expert muses, "I could say that the factor may have been down there for centuries, but that it remained uninterested in us so long as we did not disturb it by probing into its environment."[48] The narrative turns back to debating whether "the factor" could be intelligent. Alastair Bocker, "an eminent geographer," submits a memorandum, arguing that "conditions, such as pressure, temperature, perpetual darkness, etc. in those regions made it inconceivable that any intelligent form of life could have evolved there," concluding that the events must have been caused by "interplanetary invasion."[49] More scientific speculation ensues regarding whether the creatures are mining metals to colonize other territories. An alternate sort of "Anthropocene" is envisioned, as the intelligence of these creatures would seem to necessitate their desire to alter their environment through extraction of materials.[50] No other species are referenced here, only the humans who have altered the planet and the unnamed creatures. The cetaceans, who would, just a couple of decades later, be difficult to ignore as a highly intelligent marine taxon, are not mentioned. Moreover, "Phase One" concludes with a draft of Phyllis's script, which echoes Edward Forbes's azoic hypothesis of 1843, declaring that the darkness and pressure would make it impossible for life to exist below three hundred fathoms: "It is an eerie place, an awful, place, death's own place . . . absolutely nothing could live here. This is beyond the reach of life: the nethermost pit."[51] This sensationalist imagining of the depths as a void seems the flip side of boredom rather than an aesthetic of wonder or multispecies speculation that would reconfigure the terrain of concern.

By "Phase Two," ordinary people have heard of the ruckus, speaking in pubs about the "whatsits at the bottom of the sea" and the "Things in the frickin' Deeps, for crysake!" Finally, the term "xenobath," derived from "xenobathetic intelligences," "began to be used in print."[52] David Ketterer writes that the "undersea 'invaders' remain incomprehensible" despite various hypotheses about whether they evolved in the deeps or were an "interplanetary invasion."[53] He notes that Wyndham "had to fight very hard with his American publisher (but not, it would appear, his English one) to retain the incomprehensibility fundamental to his conception."[54] Ketterer argues that the beings are less mysterious than they would seem, at least on a metaphorical level, as they "can be understood as manifestations of a female sexuality," with their oval, egg-like shapes and their slime.[55] Thus, in "symbolic terms, the aliens' purpose is castration"—or, more pointedly, "the xenobaths are the Lorena Bobbits *[sic]* of the deep."[56] Rather than resolving the incomprehensibility of the whatsits, by locating a psychological source for their symbolism and dispelling the mystery with a cartoonish phallogocentrism, readers can remain with the bafflement,[57] speculating about, but not finally capturing, these ungendered abyssal creatures. We can take our cue from the "kraken" in the title, a term that has been used in some form for centuries to describe mysterious creatures. The kraken is usually thought to be a large octopus or giant squid, but it has also been posited to be a crab or cuttlefish, or even something that could be mistaken for an island.

Matters of evidence arise again when the reporters are sent on an expedition to document the creatures because the sponsors of the national media think "a description, some photographs, and definite evidence of the nature of the Deeps creatures was well overdue."[58] This is quite an understatement, as the creatures wage war with people who haven't the vaguest sense of what or who their opponents are. The team witnesses a "sea tank" of "dull, grey metal" moving through the square, which grows a "dome-like excrescence," a "semi-opaque substance which glittered viscously."[59] This unnamed entity—which may be a machine, a creature, or part of a life-form—is then described as a "huge, repulsive bladder" that "shuddered jellywise" and "lurch[ed]" in an "amoebic way" before bursting "into instantaneous bloom by a vast number of white cilia which rayed out in all directions."[60] If that weren't enough, the "thing that had burst" becomes a "round body no more than a couple of feet in diameter surrounded by a radiation of cilia," which were lashing out and capturing people in a tentacular fashion.[61] At the end of this scene, which is less horrific than it would seem, given the slime, fetid odor, and human abduction, the extreme bafflement caused by these confounding things that may possess an intelligence equal to that of humans is assuaged by the fact that one of the members of the expedition managed to secure a "specimen" of "one of those tentacle things." Conventional anthropocentric epistemologies and scientific practices do not comfort for long, however, as that same page brings the theory that the deep-sea things are on land; Phyllis asks whether they intended "to catch and collect people, like—well, as if they were sort of shrimping for us?" The scientist hesitantly agrees: "Clearly the primary intention was capture."[62] The characters do not linger on their potential status as specimens for another intelligence, however, as their conversation shifts back to debating what these things are, whether they are sentient, and how intelligent they could be. One conclusion is that the specimen, and by extension the various manifestations of these things, are "strange enough to baffle the experts."[63] "Phase Two" concludes with Phyllis asking Bocker if he believes they are more intelligent than humans. He responds, "They think in a quite different way—along other lines from ours."[64] Readers attuned to posthuman theory and critical animal studies may be struck by the irreducible difference of these beings.

"Phase Three" describes humans muddling through as "the enemy" melts the ice caps and floods human habitats globally, leaving only one-fifth to one-eighth of the world's population alive as the battles subside. Echoing Hugo Grotius's seventeenth-century "Mare Liberum" and Edward Forbes's azoic hypothesis, our narrator offers a toast on the final page: "Here's to empty Deeps and free seas again."[65] Note the capitalization of "Deeps," which signals its distinction and sovereignty. While the characters wish to return to the normal course of capitalist, colonialist, and extractivist enterprises that empty the seas of intelligent creatures, the narrative itself would not exist without the bathies. Indeed, the calmness and quotidian resolve permeating this ostensibly apocalyptic tale shifts readers' attention elsewhere—to the Deeps themselves, which remain a site for speculation, by readers if not always by the characters. In one of the most emotional sections of the book, however, Phyllis is wracked with anxious curiosity:

> But what are the things down there going to do then? Sometimes I dream of them lying down in those dark valleys, and sometimes they look like monstrous squids or huge slugs, other times as if they were great clouds of luminous cells hanging there in rocky chasms. I don't suppose we'll ever know what they really look like, but whatever it is there they are all the time, thinking and plotting what they can do to finish us right off so everything will be there.[66]

In keeping with the surrealist's respect for dreams, this passage expresses both an urgent desire to know what the creatures look like and a resignation to the fact that they elude visual capture. Phyllis had complained earlier of "too much geography . . . and too much oceanography, and too much bathyography: too much of all the ographies and lucky to escape ichthyology,"[67] but her dreams ironically voice a desire for ichthyology. Recall Oreskes: the ocean science in this period could have been otherwise—less militaristic oceanography and more marine biology.

Phyllis describes her dreams of the "sea bottom," imaging benthic lifeworlds, only to then place all her hope on capture, objectification, and war: "if only we could capture one and examine it we should know how to fight them."[68] The novel concludes with Phyllis asking whether the latest warfare against the bathies—deadly ultrasonic waves—has yielded any knowledge. "But have they discovered what Bathies are . . . What they look like?"[69] Cue Oreskes again, to explain how oceanography in the 1940s and 1950s was aimed at military pursuits rather than marine biology. The novel's answer echoes the repeated historical failures to capture gelatinous animals as specimens: "Not so far as I know. All Bocker said was that a lot of jelly stuff came up and went bad in the sunlight. No shape to it."[70] These dismissals reveal a lack of wonder as well as a lack of scientific information. The characters also lack speculative talents as their attempts at understanding the bathies are constricted to anthropocentric expectations regarding the shape of life itself. The novel seems to critique the sophisticates' lack of curiosity even as it declines to conjure up creaturely perspectives. The endless polite chattering about the bathies is not exactly riveting, but appropriately, it does leave the desire for the unknowable unquenched, as science, journalism, and literature become modes of inquiry in which speculation builds on fact. Moreover, the narrative's rather nonchalant tone discourages readers from demonizing deep-sea life. The novel's submersion of the marvelously bewildering abyssal life-forms within the unflappable tone and sedate pace of the narrative may encourage a transfer in which creatures from the abyss, however strange they may seem, are understood not as monstrous or abject or alien but instead as irreducibly different in their intelligence, yet to themselves, for themselves, and in their place, utterly ordinary. As Mike and Phyllis conclude the novel by pondering what it will be like to live with only "a fifth or an eighth of us left," they take the long view, pondering geological epochs in which their ancestors' hunting grounds were drowned by invading seas.[71]

In *Animal Alterity: Science Fiction and the Question of the Animal,* a comprehensive analysis of science fiction as human animal studies, Sherryl Vint argues, "In sf we can once again find ourselves confronted by the gaze of 'absolute alterity,' an other who looks back at us from its own point of view and often one whom we must acknowledge as having power comparable if not identical to our own."[72] The whatsits of *The Kraken Wakes* epitomize an absolute alterity. As beings with comparable power, who threaten to melt the polar ice caps and drown terrestrial human habitats, they dramatize the "problems" that the animal, imagined as an independently minded, powerful alien could cause for the human. Vint concludes one of her chapters, "Existing for Their Own Reasons: Animal Aliens," by noting that some of the texts she discusses "use the conflict [between humans and animal aliens] as a point of critique of humanity's xenophobia," yet "the prevalence of these figures speaks to our continuing desire to find some way to connect with other life."[73] Wyndham's novel can be read as an expression of that—in this case thwarted—desire. Moreover, the bathies constitute a "novum," to invoke Darko Suvin's term, which Istvan Csicsery-Ronay Jr. defines as "the central imaginary novelty in an sf text," which "fuses aesthetic effect with ethical and historical relevance."[74] While the bafflement of the novel does produce some aesthetic pleasure, such pleasure ultimately fizzles out into species-bereft boredom. Csicsery-Ronay argues that the novum offers "a vertiginous pleasure, more ludic than cognitive, more ecstatic than disciplinary," offering "the ecstatic sense of being freed from predetermined relations, the opening up of a familiar, fully mapped, and hence seemingly enclosed world, out from the authoritarianism of the current version of technoscientifically defined reality."[75] This framework works to a certain extent, as the "waking" kraken signals a critique of technoscientific and militaristic orientations toward human and nonhuman life. But the anthropocentrism of this reading would sublimate abyssal life, using animal alterity as an impetus for human ideological awakening, devoid of an ecstatic, vertiginous, and often surreal sense of the myriad modes of knowing and being of other species. I agree with Vint that sf as a genre has the potential to critique human exceptionalism and envision alternatives: "In sf the animal is in us and we are the animal, all continually involved in a never ending process of becoming, of imagining new ways of conceiving humans and animals, new ways of organising our social relations, new futures to inhabit."[76] The deadening shift from the vibrant scientific, aesthetic, and philosophical imaginaries of Beebe and Bostelmann to the lifeless, militarized oceans of World War II and the Cold War signal that the deadpan tone of the novel conveys a loss of interest in other species—particularly those in the depths—rather than a nonchalant boredom. The kraken wakes; the humans do not. Speculation without a sense of encounter, recognition, or aesthetic marveling—without the palpable sense of other beings as beings, in all their enigmatic weirdness or shimmering beauty—spirals into a vacant cognitive abyss.

Farmed Whales, Captured Giants, and the Ones Who Got Away

British sf author Arthur C. Clarke explains the origins of *The Deep Range* (1957): "I had just become seriously addicted to underwater exploration, and soon afterward bought my first scuba set."[77] Clarke's sense of the ocean as a realm of aesthetic pleasure and multispecies speculations appears in his 1955 book of essays, *The Coast of Coral,* which begins by indexing the wonders of coral, a "jewel" with "beauty" and "special magic," and concludes by predicting a future filled with "not merely underwater hotels, but even underwater homes," a "sanctuary where men and fish could share the water in peace, each studying the other's peculiar ways."[78] Margaret Cohen, in *The Underwater Eye,* explains how developments in scuba and underwater filming technology "created new access to a vastly expanded range of marine ecosystems, in different waters and at different depths, as well as the ability to portray them with greater clarity and with new kinds of movements, new camera angles, and in mesmerizing color."[79] She adds that the "newly accessible underwater realm was so captivatingly different that it created for some viewers an impression that it was at once poetry and fact."[80] This insight—that aquatic worlds are simultaneously poetry and fact—is consonant with my aim in this book: to trace how the aesthetic of deep-sea life surges through and interconnects science, art, and literature. Philippe Diolé, Cousteau's diving partner and cowriter, "devoted an entire chapter to the poetry of the depths," suggesting "that the reality of the underwater environment created new realms of fantasy for the imagination."[81] In that chapter of *The Undersea Adventure,* published in French in 1951 and in English in 1953, Diolé argues that poetry,

> before scientific definitions can be realized, is the least misleading language in which to conjure up conditions that are barely recognizable and which the mind cannot grasp. Because of this the poet is the precursor even of the expert in the exploration of the sea. Writers and poets are as important to its unraveling as biologists and geologists: they are intellectual weapons that strike at the heart of the mystery.[82]

Diolé's theory places the aestheticized act of poetic making before scientific definitions and practices. One of Diolé's captions to a black-and-white photo of a diver holding a fan-shaped coral, swimming toward the viewer, declares an ethically saturated version of unknowing: "The underwater world escapes the hands that clutch at it."[83] Clutching is a rough act, a mode of capture devoid of nuance, sensitivity, or reciprocity. Although the early Jacques Cousteau's brutality against marine life makes many contemporary readers and viewers wince, Diolé includes captions expressing concern for the suffering of marine animals. Captioning a portrait of a hovering octopus with limbs arrayed like a star, he reprimands, "Nothing, except the wrong kind of vanity induces a diver to bait an octopus"; captioning a fish floating oddly on the diagonal, he laments, "In the water death is a drama as painful to behold as that of a dying hind on earth"; and captioning a close-up of a ray with a metal hook in its mouth, he laments, "Fish suffer too, even though they die in silent convulsions."[84] Diolé's concern for the suffering of marine life, considered by most Westerners to be merely "seafood," was well ahead of its time. By contrast, Clarke, in *The Coast of Coral,* tells, without remorse, how he baited an octopus, his "victim wriggling at the end of the spear."[85]

Clarke's sf novel *The Deep Range* (1957) nonetheless contains glimmers of aesthetic and ethical regard for deep-sea life, albeit within a grotesque account of human domination of cetaceans. *The Deep Range* depicts a future in which plankton farms and whale ranches feed the world.[86] The middle-manager protagonist, Walter Franklin, experiences a few moments of aesthetic and ethical regard for ocean life, but he is a company man, or in current parlance a "tool," a cog in the machine of industrial scale domination of other species. After a traumatic accident in space, Franklin becomes a "warden," managing the industrialized slaughter of whales until a global Buddhist movement forces the whale farms to shift from slaughtering the whales to milking them. In both processes, the "obedient" whales, compared to cattle and "well trained dogs," line up to be milked or slaughtered.[87] Despite the narrative's sustained attention to the rather dull protagonist, the implied author guides us with explicit criticism: "Like most men of his highly materialistic era, he was intoxicated with the scientific and sociological triumphs which had irradiated the opening decades of the twenty-first century."[88] Such lofty language follows a scene of highly efficient carnage, when Franklin shows Thero, the Buddhist leader, how the whales are killed: "It had taken less than two minutes to reduce a lord of the sea to a bloody shambles which no one but an expert could have recognized." Thero responds to the carnage by stating that Buddhists believe "all creatures have a right to life."[89] While the novel critiques the treatment of the whales and other ocean creatures, it endorses the tidy, condescending, patriarchal conclusion in which Franklin's son follows in his father's footsteps, ascending into space, and Franklin is left to "shepherd" his wife and daughter and "guard" the whales, "his subjects."[90] Domestication, domination, duty. While the process of reducing living whales to bloody shambles leaves a foul aftertaste, somehow the compliance of the female counterparts to the "lords of the sea," the whales trained to be milked by human technology, provides an ethical compromise for the plot. (Never has being a vegetarian but not a vegan seemed more grotesque.) In his imagining of a biopolitical patriarchy that rules across species lines, Clarke provokes obvious ecofeminist critiques. Weirdly, for example, Franklin judges a humpback whale with her calf to be "no beauty," reducing her to an object to be used rather than an aesthetic marvel, just before a conversation calculating expected amounts of milk production.[91] By marrying Franklin and bearing his children, Indra, a marine biologist, is domesticated, but she manages to "still" perform "part-time work at the Hawaii aquarium when her household duties permitted"[92]—one confined being tending to others. Clarke, as the implied author, is often critical of Franklin, yet there are moments when the human exceptionalism and misogyny of the implied author and his protagonist cannot be distinguished.

While the novel's whales are domesticated for factory farming of milk or flesh, deep-sea life remains more elusive and aesthetic. In its fleeting depictions of deep-sea creatures, the novel betrays a longing for a time when the ocean was not ruled by Man. Such aestheticized moments provide an escape hatch for readers who wish to exit this reductive technocratic world, enticing us with ocean life as something other than seafood and sparking speculations about oceanic realms beyond anthropocentric knowledge, use, and control. Franklin makes his first appearance in the novel as he waits for his new boss; looking around the office, he swiftly "examined and dismissed the deep-sea photographs hanging on the walls."[93] His lack of curiosity and aesthetic attunement become even more apparent when he dismisses the most "eerie sound in all the world . . . the screaming of a herd of whales" by noting that "weirdness lies only in unfamiliarity."[94] A few pages later, however, Franklin glimpses a "long thin object" that was "moving quite rapidly," which he chases for two miles before giving up. His colleagues advise him never to tell reporters about the creature that got away because he'll "never live it down."[95] Indra, however, is nonplussed by the sea serpent story that Franklin brings home, asking simply, "Yes, but *which* sea serpent," explaining that there are three types: the giant eel, the oarfish, and "the one we haven't identified or even described," noting, "We just call it 'X' because people still laugh when you talk about sea serpents. The only thing that we know about it is that it undoubtedly exists, that it's extremely sly, and that it lives in deep water." Franklin responds:

> He did not like to admit that, despite all the instruments that man now used to probe the sea, despite his own continual patrolling of the depths, the ocean still held many secrets and would retain them for ages yet to come. And he knew that, though he might never see it again, he would be haunted all his life by the memory of that distant, tantalizing echo as it descended swiftly into the abyss that was its home.[96]

The ideal reader, of course, would also be haunted by the mysterious serpent and its abyss.

This experience transforms Franklin, who becomes "whale happy," a "somewhat indefinable term which might be summed up as acting more like a whale than a man in a given situation," a "form of identification" that could become "too extreme." He begins to relish the "life and wonder in the sea" and feel "a sympathy and an almost mystical reverence, of which he was half ashamed, toward the great beasts whose destiny he ruled."[97] A few pages later, however, Franklin must sort the problem of a giant squid who has been killing whales and costing them money. One of the biologists argues that Percy, the squid, is "one of the biggest animals alive on this planet," so it would be a "crime to murder him." Sadly, even the biologist is motivated by money rather than curiosity because he knows that Marineland would pay a sizable sum for this sizable creature if captured alive.[98] As Franklin descends into the abyss, hunting for Percy, he describes "half a dozen fantastic deep sea fish following him—nightmare creatures, two or three feet long, with enormous jaws and ridiculously attenuated feelers and tendrils training from their bodies." When they come upon the squid, it is described metaphorically: "A forest was walking across the seabed—a forest of writhing, serpentine trunks."[99] Despite that charming metaphor, within a novel in which whales are equated to domesticated cattle, the imposition of the forest onto the seabed reveals the lack of the poetic capability that Diolé had exulted: the ability to conjure up what the mind cannot grasp. As Franklin watches the squid wrap "his" arms around his friend's submarine, he can't help but notice that "he's going through the most beautiful color changes you can imagine—I can't begin to describe them."[100] Back at home, Indra reads giant squid passages from Melville's *Moby-Dick:* "We now gazed at the most wondrous phenomenon which the secret seas have hitherto revealed to mankind . . . an unearthly, formless, chance-like apparition of life." She concludes: "But Ahab said nothing."[101] As in *The Kraken Wakes,* the lack of an aesthetic sensibility, the absence of awe, an incurious detachment toward other beings, is a failure. With the blasé dismissal of breathtaking beings, creatures and lifeworlds are banished, stilled, forgotten.

Such a lack of aesthetic receptivity, such a lack of wonder about the being of other creatures, results in cruelties committed under the sanitized sign of science. After being bombed with narcotics, Percy, described as "five or ten tons of gelatinous body," is captured, collared, and towed under the sea. In the middle of this process, as if to signal his aquatic creaturely being, he begins to "glow a beautiful blue."[102] This aesthetic glimmer is left behind as he is reduced to being a "monstrous captive," a specimen whose immense size was no match for human technological "efficiency": "The caging of Percy had taken less than a quarter of a minute."[103] Franklin does not hesitate to perform his role as warden, but the passages focalized through his perspective reveal the cruelty. As the giant squid "investigates its prison" and touches the electrified net, Franklin notices that the squid looks at the "puny spectators with a gaze that seemed to betoken an intelligence every wit as great as theirs."[104] While the squid sees the puny spectators, is it possible for them to see him? "Their eyes ran along the hundred and more feet of flexible, sinewy strength, the countless claw-ringed suckers, the slowly pulsing jet, and the huge staring eyes of the most superbly equipped beast of prey the world had ever seen."[105] But, similar to the caged zoo animals John Berger describes, the squid "constitute[s] the living monument" to its own "disappearance."[106] Indeed, for the warden, the squid's status as a marvel is momentary as it becomes just another management problem. The squid, sea serpent, and other stunning deep-sea species are a blip on the screen of the novel itself, as curiosity, aesthetic appreciation, and ethical concern for other creatures do not thrive within the workaday world of industrialized, efficient subjugation and extraction of marine life. While the plot of *The Deep Range* follows its distinctive novum—the extension of terrestrial factory farming into the ocean—the implied author, the terminology, and the double voicing condemn these practices, but only ambivalently so, as both the author and the protagonist forge ahead with what needs to be done: slaughtering, milking, capturing giant squid, or making the middle manager of extractive industrial capitalism the center of the tale. There is space for readers to see this future world as more wretchedly brutal than it is depicted, however, and to take refuge in the marvelous depths, speculating about the mysterious and beautiful creatures of the abyss, and delighting in sea serpent X, the one who somehow got away.

Amateur Biology, Ephemeral Specimens, and Protected Gardens

Jetting ahead a decade to 1967, scientists descend in *Alvin,* the human-occupied vehicle constructed in 1964 for Woods Hole Oceanographic Institute, and spot what they call a "sea monster." If only it had gotten away! Victoria A. Kaharl, in *Water Baby: The Story of* Alvin, describes how it was “undulating like a translucent python, its long body seemingly unending.”[107] Then one of the scientists grabs it with a mechanical arm, drags it, and lets it go when another scientist wanted to use the arm to pick up rocks. While the “monster” was captured and presumably harmed, if not killed, as it was dragged along, it was not scientifically captured; it did not become a specimen; it was not studied. It was a cruel and useless encounter, careless and grotesque. Because *Alvin* was constructed for and used primarily on military-oriented expeditions rather than for biological research, such improvident episodes were apparently not infrequent. Kaharl tells of one incident in 1967 when a swordfish collided with the submersible. Sadly, “no ichthyologist or any other kind of biologist was on board to tell them about *Xiphidai gladius* or to appreciate its beauty even in death. The creature, which was eight feet long and weighed 200 pounds, was in a sorry state. At the enormous pressure change from 2000 feet to sea level, the swordfish erupted inside, spewing its innards on the deck.”[108] Kaharl keeps her narration light, but echoing Oreskes, it is clear that specimen capture and marine biology generally, were not a priority: “There was plenty of swordfish to go around for supper which was easily the best meal ever served aboard Lulu. WHOI’s [Woods Hole Oceanographic Institute’s] front office was piqued to learn that its own oceanographers had eaten ‘the specimen.’ ”[109] Whereas Miguel Covarrubias’s caricature of William Beebe’s partaking of the specimen suggests an eccentric intimacy, here, consumption is just another mode of disregard. Other lively creatures hauled out of the depths met similarly messy fates. Kaharl tells of how, for example, during the dives into the newly discovered deep-sea vents, the Galápagos team “wasn’t prepared to handle the hundreds of retrieved animals. They didn’t know how to pickle them anyway; on a geology–chemistry cruise there was no need to bring formaldehyde along. . . . Neither were there any jars to hold the animals.”[110] As they try to make do with Tupperware and plastic wrap, biologists radio the team, urgently instructing them on basic practices to preserve the specimens. Robert Ballard, a former naval officer and marine geologist who participated in over a hundred deep-sea dives at WHOI, recalls, “Our thoughts had been so far from biology that we brought no preserving fluid along, except for a small amount of formaldehyde. Some of these amazing specimens therefore, made the trip back to shore immersed in our strongest alcohol: Russian vodka we had purchased in Panama."[111] While the transformation of living creatures to dead specimens (i.e., killing) is hardly something that an animal ethics would endorse, the careless, grotesque capture of these creatures and the lack of concern for the practices of marine biology that would enable their deaths to at least yield information that could support future conservation efforts remain troubling. These scenes, as well as the fictional capture of the giant squid in *The Deep Range*—collared and dragged and caged—are revolting, as the organism loses its coherence when it is hastily rendered[112] into material for profit, or dinner. Such corporeal vulnerability can inspire an ethical sense of connection between human and nonhuman bodies, but it can also result in recoiling from the abjected state of "the animal" as flesh. An animal ethics would embrace an insurgent, interconnected vulnerability, an orientation toward respect, care, and concern.

Jacob Darwin Hamblin notes in *Oceanographers and the Cold War* that "the most significant change in international marine science in the 1970s was the importance placed on environmental issues, sparked by devastating oil spills in the late 1960s and controversies over marine pollution of various kinds."[113] Glimmers of a more ethical scientific practice appear in the 1979 National Geographic Special, *Dive to the Edge of Creation,* documenting an expedition to the deep-sea vents. Biologists remark, for example, that the red color in tubeworms is produced by blood, just like the blood that runs through humans. Moreover, various scientists on this expedition express great interest and concern for the fragile peach-colored gelatinous ball they call a deep-sea "dandelion," hoping the lovely specimen they managed to bring up from the vents will not disintegrate. They carefully place the intact dandelion into its own jar. One of the fourteen biologists aboard remarks, as he injects it with killing/preserving fluid: "Hate to kill something that is pretty as that." The narrator remarks, "Within an hour the fragile dandelion disintegrates; apparently mankind will gaze on this fragile creature only in pictures or 8,000 feet down."[114] The deep-sea species that cannot withstand the dramatic change in temperature and pressure when brought to the surface testify that they are not of this airy world. To render these lives into stable, distinct specimens means transforming them into something antithetical to their very being because, as it turns out, the dandelion happens to be a siphonophore—

not one individual being but rather a gelatinous colony of zooids that link up with each other to survive.

Although the National Geographic Special, produced for popular consumption, features scientists marveling at deep-sea life discovered at the vents, the predominant view in the serious scientific workplace was that such aesthetic responses needed to be contained and discarded. Ballard summarizes the biologists' descents in the 1970s: "Thus biologists quickly got over their 'oh's and ah's,' as Beebe once put it—the early, unproductive phase of sheer amazement. Within a few years after Hessler had made the first science dive in *Alvin,* the whole field of benthic biology was rapidly moving into a far more quantitative phase."[115] Beebe, as chapter 1 documents, did not consider the aesthetic and the emotional to be something that impaired science; nor did he exult the quantitative above the qualitative. The fact that aesthetic responses to the depths become a problem testifies to their potency, their ability to distract from a narrow, productive focus. By 1976, *Alvin* teams had to compensate for the exhilarating wonder the depths aroused. "Newcomers were urged to make at least one dive before the expedition. This preparatory 'gee-whiz dive,' as the scientists called it, was so filled with ooo-ing and ah-ing that serious work was largely left undone and the tape-recorded observations and written notes made by the dazzled scientists made little sense."[116] The deep sea unmoors human visitors, immersing them in a state of wonder where they could hardly think straight as reasoning dissolves into marveling. This aesthetic response flows across divides between experts and ordinary people because its immediacy requires no prior knowledge. Kaharl opens one chapter with an epigraph from McManis, the pilot, who describes what he saw during an oceanographic descent in 1967: "Then I seen this monster or somethin'. . . . I turned around sharply and it's gone. Kind of shook me up. Could it be possible? This was a living creature. I seen at least 40 or 50 foot of it. I swear to God. Like the other day, a ten-foot worm. The first few feet, it looked like a kind of caterpillar with fuzz on it and the next few feet, it looked like a bunch of Ping-Pong balls glued together." McManis delights in the fact that the biologist knows no more than he does about this fish: "the amazing thing is I ask him what type of fish this is and he's as much in the dark as I am."[117] An exuberant democratization of knowledge is sparked by the marvelous being, as ordinary people can encounter unknown species without that experience being channeled, controlled, or delimited by scientific categories and expertise. Never mind that the idea of the unknown is well worn, and that the Ping-Pong ball creature could be placed next to, say, an upside-down urinal in the now tired traditions of the aesthetics of novelty and disruption. Of course, nothing is new under the sun, yet the cynical academic stance that would deflate exclamations of wonder fails to suppress impassioned responses to the sight of as-yet-unseen creatures dwelling in sunless worlds.

Notwithstanding Robert Ballard's disparagement of "sheer amazement" as an "early, unproductive phase," his personal account of the hydrothermal vent discoveries tells a different tale, one in which wonder and excitement transform those who are "not biologists" into "amateur biologists." During the Galápagos hydrothermal vent descents in 1977, they find themselves practicing biological capture and speculation. "We couldn't help but wonder what so many animals were doing at that depth, in that eternal darkness. Where was their food supply? What source of energy could they possibly be tapping?" Members of the team stared "in amazement" at giant clams, "weird stuff growing on the bottom, like dandelions," and "the wormy stuff attached to rocks."[118] In "The Garden of Eden," they witness the tubeworms, "the strangest creature we had ever seen," and lo and behold, "midway through our cruise, we all became amateur biologists. And why not? We had discovered a new realm of life. We began to speculate that the animals in these deep-sea oases got their energy and nutrients not from the sun and photosynthesis but from the earth's hot volcanic depths. We felt as if we had glimpsed unknown, alien life on a new world, or at least an alternative version of life on our own." A team member recalls exuberantly, "We were dancing off the walls!"[119] Because Ballard's book was published thirty years later, we might consider whether his recollection of these marine biological speculations are indebted to the subsequent work of actual marine biologists. Nonetheless, the shift from the careful segregation of aesthetic and emotional responses to an ecstatic transformation into being amateur marine biologists testifies to the potency of creaturely encounters, as well as the confluence between science and the aesthetic.

While an aesthetic response to deep-sea life may seem a flimsy, willowy, and ineffectual mode of encounter, it provides friction against the brutal and unthinking objectification and extractivism endemic to ideologies of human exceptionalism and domination, and in so doing, it opens possibilities for the environmental protection of abyssal ecologies. Many biologists are driven not only by intellectual curiosity, which itself cannot be segregated from wonder, but by an attachment to—and, dare we say, love for—the creatures they study.[120] Imagine the excitement of Bob Hessler and Howard Sanders, marine biologists who had been dredging samples out of the deep sea, when they descended in *Alvin* a hundred miles off of Cape Cod: "Their plan was to take photographs from *Alvin,* collect animals and above all, they were going to look—to finally see what they had studied for seven years in the vicarious fashion of blind sampling and educated guesswork."[121] Kaharl's account overflows with the scientists' curiosity, excitement, and wonder. Sanders says, "There are so many things to see, all these things, new phenomena, and that is all that matters; you finally have no fear." Hessler is not only amazed by the abundance of fish but also by the sheer beauty of particular creatures: "Oh! . . . A beautiful deep-sea arrow worm with large gonads just passed the porthole."[122] Later, Kaharl presents a litany of biologists testifying to the importance of being immersed within the lifeworlds they study. WHOI biologist Lauren Mullineaux attests that her "first dive changed [her] whole ideas about what the deep sea was like." Biologist Craig Smith declares, "It's unbelievable just how much can be gained by going to the bottom of the ocean and just thinking about what you are seeing."[123] While these statements may be vague, they suggest the intermixing of the aesthetic, the emotional, and the cognitive, an immersion into the field of study that propels knowledge that is neither dry nor detached. Echoing the ideas that psychedelics act as portals, Alice Alldredge says *Alvin* "opened a door": "It allowed me to collect and really observe, really see the material. And that allows me to conceptualize the deep sea. It may not sound like much, but how we conceptualize is directly related to the kinds of questions we ask." Alldredge's findings on the constitution of marine snow, with Mary Silver, even have an *Alice in Wonderland* quality: "there were actually large viable populations growing on these particles . . . They weren't just debris. . . . They were homes."[124] While none of these biologists uses the words "beauty" or "aesthetic," the revelatory quality of simply seeing and thinking about what one is seeing—something so basic yet so impossible to describe—as well as the emphasis on conceptualization places the scientific process within an uncontained realm of perceiving, sensing, feeling, being affected, thinking, questioning, and conjuring up concepts from within a state of curiosity, wonder, and awe.

Given that *Alvin* and its expeditions were primarily in the service of the U.S. military, the fact that some biologists participated in the vent dives and a few more had access to the specimens is a fortunate, if rather accidental, turn of events. Oreskes argues that "both the idea of *Alvin* and its technical realization were rooted in military needs, as were the instrument specifications and delivery deadlines," adding that Kaharl's "popular history" of *Alvin* "maintained the myth that the military connection was incidental."[125] While Oreskes presents a persuasive case that U.S. military funding distorted ocean research, resulting in the neglect of marine biology as well as research pertaining to climate change, and while the discovery of deep-sea vents is one of the most stunning episodes in this argument because of the sheer scientific importance of the vents as an ecosystem abundant with life, including what would turn out to be chemosynthetic (not photosynthetic) life, her disparagement of aesthetic responses to the vents fails to recognize the potential for such responses to inspire alternatives to the narrow, hardheaded naval agenda. Before quoting extensively from geologist Jerry Van Andel's personal diary, which details the beauty of the "coral gardens, pink and gold anemones," and other life within this "little paradise in the . . . sea floor desert," Oreskes disparages his description: "What they found has been described as 'one of the most exciting developments since the beginning of oceanography.' Yet the initial observations of these communities read less like a eureka moment and more like a page out of Henry David Thoreau."[126] It isn't exactly clear what Oreskes is insinuating here, except that, consonant with her overarching critique, these descents should have included more biologists who may have produced more "eureka moments." Quite true. Moments of sudden insight, however, that lead to crystalizing clarity are only one mode of encounter and understanding, and perhaps not even the most productive, in terms of reckoning with utterly new lifeworlds. The Thoreau reference denigrates the aesthetic response, containing it within the realm of literature, not science, even though, as it turns out, Thoreau was not only deeply engaged with the science of his time but also read Humboldt and engaged in ecological observation.

Another episode in Oreskes's incisive and invaluable volume, however, demonstrates how the staggering discovery of chemosynthetic vent life, along with other accidental discoveries of abundant life in the depths, was ignored or forgotten by those with other agendas. Oreskes narrates how, in 1991, Woods Hole hosted a conference, "The Abyssal Ocean Option for Future Waste Management," featured in a front-page article in the *New York Times,* which included a description of the abyssal ocean as "the planet's most useless real estate."[127] Oreskes cites ecologist George Woodwell's opposition to the plan to dump "one million tons of sewage sludge" three hundred miles offshore in the Atlantic, despite the Ocean Dumping Ban Act of 1988: "The continued revelations of the *Alvin* and its successors in exploring benthic fauna . . . constitute one of the most exciting and provocative frontiers of science. . . . The proposal [suggests] that we do not take seriously our own discoveries about the oceans and flies in the face of what virtually every citizen has learned from us to accept as elementary reality.' "[128] Competing depictions of the deep seas, from fairyland to wasteland—an astounding realm of biodiversity to an empty expanse for dumping—underscore the potential impacts of cultural tropes on even the most remote biomes.

Fortunately, Oreskes reports, "by the mid-1990s, seabed waste disposal—radioactive or otherwise—was a moribund, if not deceased, idea."[129] This incident, however, along with recognition of other ongoing threats to deep-sea ecologies, and the frustration with paltry funding for ocean research, provokes some marine scientists to openly advocate for deep-sea conservation by stressing the beauty of the deep sea. Cindy Lee Van Dover was a graduate student at Woods Hole, working with Fred Grassle, processing deep-sea specimens. Desperate to descend in *Alvin* herself, she became the first female deep-sea pilot in 1989.[130] Kaharl documents how sexism permeated the *Alvin* expeditions. Not only were there juvenile outbursts (such as when someone said the ocean floor looked like "acres of breasts" he "would love to walk all over . . . barefoot"—Kaharl adds "and so on," suggesting the tiresome frequency of such comments) but, worse, female scientists were also excluded from the expeditions, were denied safe restrooms and sleeping spaces, and were subjected to harassment and even attempted rape.[131] Van Dover tells of the many ways the team undermined her confidence and impeded her efforts to become a pilot.[132]

But become a pilot she did, as well as a deep-sea marine biologist and vent expert.

In 1996, Van Dover published *Deep-Ocean Journeys: Discovering New Life at the Bottom of the Sea,* which presents an unabashedly aesthetic and environmentalist vision of vent life. Van Dover begins by explaining she has a "scientist's perspective of the deep sea" yet is also "at heart" an "amateur naturalist, quick to delight in the unusual nature of a worm, the oddities of a shrimp, the peculiarities of a rock."[133] Her aestheticized appreciation for benthic life reaches a crescendo in one particular passage within the chapter "A Chorus of Tubeworms":

> There is a singular place on the East Pacific Rise, in the Venture Hydrothermal Fields, where I have driven the submersible along the surreal landscape at the base of the jagged edge of the shallow axial valley and seen looming 8 meters tall in front of me a tower of tubeworms. The worms are so tightly packed that the basalt pillar they colonize is invisible. They look cultivated, each worm a prize specimen arranged in a formal garden just so to present a uniform face of red plumes about the pole. Because it is so unique, this pillar has become by informal consensus a sanctuary where scientists may go to look but not touch. It won't be long, a few years probably, before the warm water that feeds the tubeworms of the pillar ceases to flow and the magic will disappear. Maybe only a dozen people will ever see it.[134]

The "surreal landscape" shifts to the trope of the "formal garden," with "prize specimens." The aesthetic recognition of the singularity of the place transforms it into a site to be protected, even revered as a "sanctuary." The worms merely look, or seem, cultivated, but it is the mode of regard that cultivates an environmental vision. Van Dover's verbal depiction is complemented by a simple black-and-white illustration on the previous page, which, like some of Beebe's black-and-white photos, fails to convey the beauty of this place. Van Dover's environmental ethic cannot be separated from an aesthetic regard, a framing that encourages an orientation toward protection.

The next paragraph swerves: "Gathering live tubeworms is an art. To be of much use to biologists, the worms in their tubes must be carefully plucked and stowed in insulated containers for the long ascent to the surface."[135] Shifting from the beauty of the scene to the scientific art of specimen collection, which emphasizes use value rather than aesthetic value, may make readers flinch as we empathize with the captured tubeworms. A few paragraphs later the account swerves again as Van Dover confesses her own desires for sensual contact with the tubeworms: "I wish I could get out of the submersible, not to collect tubeworms, but simply to run my hand along their tubes and feel the gradient of warmth in the water surrounding them."[136] Seductive within their own element, tubeworms lose their phallic erotic appeal when removed from their benthic habitats: "There is nothing obviously glorious about a tubeworm on deck. Pulled out of its tube and drained of its color, the worm is decidedly flaccid and unattractive. The vibrancy of the live worm on the seafloor is lost; in the atmosphere of the sea surface the plume loses its vitality and pales."[137] While scientific objectivity, as Lorraine Daston and Peter Galison argue, fears the subject, Van Dover intrepidly ricochets between an aestheticizing, even sacralizing recognition that would protect this particular group of tubeworms, to the cold scientific practices of specimen collection, to an ecoerotic desire to touch the tubeworms, to the blunt realism of the flaccid tubeworm on deck that fails to attract. Interestingly, Van Dover chooses not to smooth over the ragged contrasts between these orientations, revealing the rifts within objectifying scientific practices of capture and the aesthetic, sensual, appreciation of other species. It isn't difficult to find an environmental ethic here, however, as "the vibrancy of the live worm on the seafloor is lost" when it becomes a specimen, underscoring that the vivid aesthetic value of particular species cannot be separated from their lives as they are lived within their own habitats. Van Dover's epilogue calls for the protection of the seafloor, noting the "shame at the desecration of something so beautiful" when she saw "plastic bags of trash" that had settled on "black branch coral nippled with minute fleshy polyps," predicting that eventually "we will turn vents into tourist attractions until we kill off all the tubeworms."[138] Valuing species or sites primarily because of their aesthetics has limitations in terms of ecology and conservation, obviously, but as a counterpoint to the deadening objectification of the scientific specimen or capitalist extractivism that demolishes ocean habits with utter disregard (literally without even seeing or recording these nonevents), aesthetic recognition can spark an orientation toward respect, curiosity, and concern. Van Dover's aesthetic appreciation for the tubeworms circulates as self-evident; she seems to believe in the obvious appeal of the appealing, a current with a long history in American conservation, which she extends to the seafloor. Indeed, Van Dover has explicitly argued for the conservation of the vent sites: "We have the opportunity now to put in place conservation measures that ensure these ecosystems are sustained and celebrated into the future as living libraries where we can continue to learn about Earth processes and biodiversity."[139]

Abyssal Monsters at the End of the Century

Peter Watts's sf novel *Starfish* (1999), set primarily underwater in the near future at the Juan de Fuca Ridge, portrays the psychological dimensions of living in an environmental and political dystopia, which entangles deep-sea creatures with modified humans, aligning a posthuman monstrosity with glimpses of a faint aesthetic.[140] Fictionalizing Van Dover's fear that the vent gardens would become a ravaged tourist destination, *Starfish* begins with the acerbic voice of Seabed Safari tour guide, annoyed by his chattering "cargo," sardonically wishing they'd be dumbstruck by an abrasive je ne sais quoi:

> The abyss should shut you up. Sunlight hasn't touched these waters for a million years. Atmospheres accumulate by the hundreds here, the trenches could swallow a dozen Everests without burping. They say life itself got started in the deep sea. Maybe. It can't have been an easy birth, judging by the life that remains—monstrous things, twisted into nightmare shapes by lightless pressure and sheer chronic starvation.[141]

The voice-over for the hackneyed video he plays for the tourists states, "We have always peopled the sea with monsters," mentioning "giant squids attacking lifeboats during the Second World War," a "litany of deepwater nasties," and the warning that even though most fish in the depths are small, on the Juan de Fuca Ridge, "Here there be dragons."[142]

The packaging of deep-sea life as monstrous is introduced as archaic, sensationalistic, and commercialized, but when it shifts to the protagonist Lenie, one of the "rifters" who has been modified to survive three kilometers down in "Beebe Station," we learn that indeed, "the rift is full of monsters" with "elastic bellies" and "unhingeable jaws."[143] Moreover, the depths have been literally "peopled" with monsters, as the rifters have been modified with the genes of deepwater fish such as the unappealingly named "rattails."[144] Clare Wall sorts three different versions of posthuman monstrosity in the novel, which "contest the liberal humanist narratives we attach to subjectivity, evolution and technological progress," underscoring how the environment of the deep sea itself transforms the characters.[145] I appreciate her attention to the effects of the material environment on posthuman transformation, a topic I discuss in *Bodily Natures.* My focus here, however, will be the twisted relations between rifter and deep-sea creature, as well as the fleeting aesthetic recognitions within the novel. When Fischer, a fellow rifter, watches Leni put on leggings that seem alive, the scene directly references an amoeba, a transfiguring microscopic organism, but more subtly incorporates the black swallower, an iconic deep-sea creature, into the scene: "Now he watches it swallowing Lenie's body like some slick black amoeba but she's showing through underneath, black ice instead of white but still the most beautiful creature he's ever seen."[146] Scopophilia, oral pleasure, and perhaps bestiality converge in this focalization as her postnatural body is swallowed by clothing that signals an engulfment by abyssal life as well as technologically enhanced regimes of social domination. When Lenie repels Fischer's subsequent unwanted touch, she repulses his framing of her as a "beautiful creature," confirming his characterization of her as self-contained, cold, unyielding "ice." Black ice, which can be treacherously deceptive, warns of the dangers lurking in the abyss. Yet since the metaphor focuses on her skin, it also suggests that Fischer's desire is enhanced by a sense of "danger," emanating from racist formations that elevate white beauty standards while eroticizing blackness: "black ice instead of white but still the most beautiful creature."[147] The postnatural corporeality of the submerged rifters is intertwined with deep-sea life in a mode of transcorporeality that denies human exceptionalism within these submerged, enmeshed zones of being. Watts refuses to externalize "nature," even that of the depths. As he insists in an interview, "We *are* nature. The problem with Humans is not that we've isolated ourselves from Nature—it's that we *embody* nature, cranked to the nth degree."[148]

Set against the novel's unflinchingly grim perspective, a deep-sea aesthetic glimmers, all the more noticeably, yet also merges with the monstrous. Watching the bioluminescent "stars" of the sparkling abyss, Lenie thinks, "They are so beautiful." Another rifter confesses to Lenie that he assumed "this place would be a real shithole," but "it's really kind of well, beautiful, in a way. Even the monsters, once you get to know 'em. We're all beautiful."[149] Lenie shifts from the aesthetic to the epistemological in a ragged reply, stating that he couldn't have known what it would be like in the rift because "nothing down here's computable." Then an "alien creature looks down at her and shrugs. 'Computable? Probably not. But *knowable—*' "[150] Such musings are abruptly abandoned because they must get to work. The aesthetic is a surprise, a tentative if not shameful experience, and an enticement to experience kinship across species lines, as "we" are all monstrously beautiful. The "alien creature" references tropes of deep-sea life but also refers to the modified human, Acton, who is alien to the abyssal environment, and himself a strangely modified creature. Tangling up the (post)human with the figure of the alien, Watts refuses the sensationalist capture of deep-sea life. The horror film trope of aliens in the deep was in play before 1997, when William Broad published *The Universe Below: Discovering the Secrets of the Deep Sea.* Writing of the *Alvin* dives that stumbled onto vent life, Broad writes, "the scientists were dealing with a world that was truly alien," later describing "giant globules of lava" as an "alien material that had been twisted into an alien landscape."[151] He describes the "revelation" of chemosynthesis in terms that could fuel a horror film: "we and all the other light-eaters of Earth shared our planet with an alien horde that thrived in total darkness."[152] Such renditions of the alien alienate the depths from the transcendent, contained human who explores, discovers, maps, knows, and exploits an externalized world. Despite its monstrous natures, *Starfish* imagines more complicated relations between posthumans and deep-sea life, akin to Stefan Helmreich's conclusion to *Alien Ocean,* in which he explains that the "links between the scale of human bodies and ecologies become baroque, spatially and temporally. The bacteria that inhabit our bodies do not simply mirror the bacteria that inhabit the sea—as might brine in our blood. This is not human nature reflecting ocean nature. It is an entanglement of natures, an intimacy with the alien."[153]

Helmreich concludes not with a "dissolution of humanity" but rather "the saturation of human nature by other natures."[154]

The brutal capitalist domination of human, posthuman, and nonhuman life that Watts presents does not foster fairylands. While there are moments when characters marvel in the beauty of the Juan de Fuca Ridge, such as when Fischer describes the beautiful site where "rocks shine green, and others where rocks shine blue," like a "vid of the northern lights he once saw, only sharper and brighter . . . a brush fire in emeralds,"[155] a cynical mood dominates the novel, dispersing a deep rage against the entangled forces of science, capitalism, extractivism, and the state.[156] Even as the rifters in *Starfish* become weirdly attached to their abyssal surroundings and enjoy moments of aesthetic wonder, a creepy, even sinister atmosphere pervades the novel as modified humans are exploited and become metonymically entangled with monsters: "He's brought a monster inside with him. It's an anglerfish, almost two meters long, a jellylike bag of flesh with teeth half the length of Clarke's forearm. It lies quivering on the deck, its insides exploded through its own mouth in the near vacuum of Beebe's sea-level atmosphere. Dozens of miniature tails sprout everywhere from its body."[157] Life becomes tortured and tortuous, from within the characters' psyches and across the expanse of the seas. The reader is immersed within the abyss; the dystopian biopolitical world Watts creates offers no comforting refuge. Moreover, the beauty that is present has already been diminished by human impacts. One character, Joel Kita, looks through the dorsal port as they descend through the deep scattering layer, observing, "Pretty thin stuff compared to the old days, he's been told, but still impressive. Glowing siphonophores and flashing fish and all. Still beautiful."[158] Then a "tiny monster" with "glowing photophores" appears, but it is so small that it doesn't "make a sound" when it hits the port—an anticlimactic, sad encounter. Jarvis identifies the "monster" as a "viperfish," telling Joel about a guy named Beebe who "swore he saw one that was over two meters long." When Jarvis holds out hope that "maybe the deep sea really *is* teeming with giant monsters," Joel informs him that it is not. This scene hints that the twentieth century has ravaged deep-sea life; whatever Beebe witnessed may no longer exist. Species are less abundant, less diverse, and smaller. If only monsters still graced the seas!

The rare aesthetic glimmers in *Starfish* evoke both the desire to imagine other worlds rich with marvelous, shimmering species who have not been forced into service and the bleak recognition of the immense reach of capitalist domination. While the characters in *The Deep Range* are all too human, the posthuman rifters are enmeshed with abyssal life as well as technology, immersed in the depths and entangled in biopolitical landscapes that stretch to the benthos. The rifters are immersed in an Anthropocene that does not spare the human, as such, from material transformation. Rifters are manufactured specimens for the machinations of the powers above them, laborers engineered for benthic environments—but also, still, people receptive to other species and the diminished beauty of the depths. Already intimate with aliens, they do not indulge in bored, sophisticated ruminations on krakens or other whatsits; instead, they crave glimpses of beauty in the deep. Such aesthetic moments are not a luxury, not a commodity, but a vital and sustaining orientation toward the possibility of vibrant, wild, and dazzling life somehow persisting, in spite of it all.

◈ 3 ◈
Counting and Framing

The Census of Marine Life

> Perhaps the ugliest of our impulses, to shove the sublime through a pinhole.
>
> —Claire Vaye Watkins, *Gold Fame Citrus*

> There is no doubt that beauty . . . has been a major victim of the Anthropocene.
>
> —Arturo Escobar, *Designs for the Pluriverse*

Novelist Claire Vaye Watkins condemns the brutal impulse to "shove the sublime through a pinhole," smashing what is vast and awe-inspiring into something impotent and diminutive. Such an "ugly" impulse surging through anthropocentrism, extractivism, neocolonialist explorations, and the commodification of something that used to be called nature would seem to be an antiaesthetic of domination, as the shoving betrays a lack of regard for what is forced through. Referencing antiquated cameras, the pinhole evokes the history of photography, suggesting that mediation, however quaint, may be complicit with harm. Arturo Escobar's contention that beauty itself has been a "major victim of the Anthropocene" echoes Watkins.[1] Although every sort of aesthetic has its devotees—brutalist architecture stoutly standing as a case in point—the already iconic visualizations of the Anthropocene, with their lofty, distancing perspectives,[2] which emphasize how vibrant ecosystems and habitats have been flattened and homogenized into mines, factories, industrial feedlots, agricultural expanses, cities, and highways, bear witness to Escobar's elegy. The capture, mediation, and scale shifting invoked by Watkins's discomforting phrase "shoving the sublime through a pinhole" may nonetheless be vital for extending environmental concern down to the deep seas, as photographs of stunning life-forms, metonyms of their abyssal habitats, circulate as enticements to care.

The turn of the twenty-first century has seen an astounding outpouring of dazzling photographs and videos of deep-sea life, flowing through social media, magazines, coffee-table books, nature films, scientific and conservation-oriented websites, TED talks, and more. Charismatic deep-sea creatures such as the dumbo octopus, a multiplicity of jellies, and constellations of bioluminescent organisms take up residence in popular imaginaries of what was once considered a lifeless abyss. Thinking with visual captures of abyssal organisms as they proliferate across the turn of the twenty-first century suggests a reflexive Anthropocene aesthetic, underscoring the tenuous networks of capture and mediation that render an unknown creature from the depths into a scientific specimen, something studied by scientists, photographed, and presented to the public. I use the contested term "Anthropocene" here to reference how the world has been profoundly altered by particular groups of humans in assemblages with technologies, economic systems, and elemental agencies. This chapter considers the accelerated period of deep-sea ocean exploration, from the late twentieth to early twenty-first centuries, characterized by excitement about descents into ever deeper zones, the discovery of heretofore unknown creatures, and their circulation on a digital stage in which photographs, videos, and narratives flow from science (always embedded in economic, social, and political systems) to the screens of publics. Some of those images are gathered into coffee-table books, which, weirdly, create a sense of encounter with the abyss—seemingly right there—in one's hands.

This paradoxical, anachronistic moment of discovery in the oceans, when unknown species are being found, photographed, and genetically barcoded at a rapid pace, happens in the midst of the unthinkable loss that is the sixth mass extinction. Triumphant narratives about newly found deep-sea life can be read as the last gasp of colonialist exploration, which deflects attention from the enormity of the capitalist plunder of the oceans while spotlighting the heroic white Western explorer and his technological marvels. Prevalent popular discourses of the deep ocean as the last frontier or wilderness grate against decades of scholarship in environmental and Indigenous studies that comprehensively critiques those paradigms. While the colossal big science project of the Census of Marine Life assumes a universalized global audience and aligns itself with the continuing cultural potency of the discourses of exploration and discovery, it also simultaneously crafts an environmentally oriented epistemology and aesthetic. The Census saturates its ambitious scientific aim to count every creature in the sea with a regard not only for the unknown but also for the unknowable, framing scientific knowledge practices within something akin to a posthumanist and environmental epistemological ethics that underscores the limits of human knowledge, especially in the deep ocean. Moreover, some of the more stylized Census photographs populate the abyss with beings that seem to call for recognition and response. My fascination with the extraordinary decade-long Census of Marine Life, particularly in terms of how it spotlighted aesthetically stylized images of abyssal organisms, inspired this book. Sadly, the fascinating, gorgeous images and videos of the Census of Marine Life website are no longer available, partly due to the rapid obsolescence of digital platforms. This is ironic given the slow temporalities of benthic life and the historical refrain that the deep sea, with its living fossils, has its own tempo. While readers of this book will probably not be able to navigate the now-defunct Census of Marine Life website, the Census coffee-table books, as well as Claire Nouvian's stunning volume, *The Deep,* may still be at hand.

Anachronistic Explorations in "Alien" Seas

While the most ethical scientific practice regarding deep-sea life would, in an ideal world, be to decline to disrupt these zones with vehicles, specimen-collecting devices, and cameras, which, together, bring light, noise, the stirring up of sediment, and the brutal transformation of living beings into dead specimens, the enormity of anthropogenic harms that deep-sea life is already suffering demands biological and ecological scientific research as a basis for and impetus to environmental protection. Halting biological and ecological scientific research in the oceans would repeat the manufactured ignorance I discussed in chapter 2, in which marine biology was cast aside, throughout the bulk of the twentieth century, in favor of a narrow set of scientific questions that supported the military.[3] That ignorance no doubt cloaked the accelerating harm to deep-sea life and marine life more generally in convenient invisibility. The militaristic disregard and abuse of ocean life continues into the twenty-first century, accompanied by the horrific scale and ever deeper disruptions of industrialized fishing and mining, along with the alteration of the ocean's temperature and alkalinity, as well as the deluge of radioactive, plastic, chemical, and biological forms of pollution.

Another sort of manufactured ignorance is produced when the abyss is nostalgically cast as an alien territory. The term "alien" is complicated when it comes to marine science studies and the blue humanities. Stefan Helmreich, in his extraordinarily rich study, *Alien Ocean: Anthropological Voyages in Microbial Seas,* explains his concept: "I use the phrase *'alien ocean'* both to diagnose a scientific, social, and cultural imagination about the sea I observed in my anthropological studies and to suggest the limits of representing the sea, for both oceanographers and social scientists. After all, the alien, as countless science fiction films have instructed us, is often a fugitive trace, a constellation of uncertain evidence in motion."[4] This striking trope of the "fugitive trace" is consonant with a new materialist ontoepistemology that traces material agencies as well as an environmental ethics that stresses the limits of human knowledge systems. Turning from microbes to deep-sea creatures, however, it becomes apparent that the popular trope of the alien has been articulated with less environmentally oriented ideological constellations. Helmreich envisions an "entanglement of natures, an intimacy with the alien,"[5] an ontological relationality. The prevalent articulations of the alien with the deep sea specifically, however, imagine it as a radical alterity, which conceals the long reach of anthropogenic harms as well as the scientific understanding of abyssal habitats and species. As the introduction briefly discussed, the anachronistic discourse of heroic deep-sea discovery revels in the innocence afforded by realms that are uninhabited by humans, constructing a convenient *aqua nullius* seemingly untouched by colonial histories of exploration saturated with genocide, the transatlantic slave trade, and industrialized plunder. The descents of "explorers" such as James Cameron and Victor Vescovo demonstrate Tiffany Lethabo King's argument that the "production of the White conquistador-settler is an ongoing process of violent autopoesis that must be continually rewritten and revised."[6]

Reports of James Cameron's 2012 descent into the deepest place on earth cast the Mariana Trench as an empty space of nothingness, an alien world, another planet. Cameron reflects: "I just sat there looking out the window, looking at this barren, desolate lunar plain, appreciating," but feeling "complete isolation from all of humanity."[7] Deep-sea biologist Alan Jamieson in his book *The Hadal Zone: Life in the Deepest Oceans,* commenting on Cameron's 2012 descent to Challenger Deep, the deepest place on earth, argues, with dismay, that "a high profile such as this can detract from years of scientific research." Indeed, Jamieson explains that Cameron's solo descent "gave the impression that Challenger Deep was unexplored and poorly understood, despite the wealth of already published research from the same area spanning over 30 years."[8] He warns that "sensationalist exploration stories" may "raise awareness" (he uses scare quotes), but "there is a danger of hindering rather than responsibly promoting dissemination to the public about scientific fact, albeit often less publicly digestible."[9] Cameron nonetheless penned a short blurb for the book in which this critique appears. Jamieson, in his TED talk, "Meet the Mysterious 'Monsters' of the Deep Sea," dispels the myths that deep-sea life is scary by showing the adorable dumbo octopus, along with other nonthreatening, diminutive, and not at all weird animals, with the explicit intention of replacing fear of the depths with concern.[10] While sensational stories of exploration may detract from scientific fact, casting abyssal life as alien to this planet or as subordinate to imaginary life in outer space may be even more pernicious. Such mobilizations of alien tropes are not speculative engagements with scientific captures but a hierarchical structuring that devalues deep-sea species by disregarding their being and their lifeworlds.

Cameron's film *The Abyss,* for example, concludes by transforming gelatinous deep-sea creatures into space aliens. While the science fiction genre of *The Abyss* makes the transformation from marine life to space alien less jolting, Cameron's *Aliens of the Deep* is a *documentary* about deep ocean science, so the arrival of a space alien at the conclusion is quite a surprise. (Wait, what?) *Aliens of the Deep* depicts Cameron, along with marine biologists, astrobiologists, other scientists, an astronaut, and Russian Mir space station pilots exploring the ocean via submersible vehicles and rover cameras. Strangely, the viewer learns little about abyssal life and even less about ocean conservation. Dijanna Figueroa, a marine animal scientist (and a Black woman) does mention climate change, but that fleeting moment is drowned out by the rest of the film in which exploration is promoted as an end in itself. When Cameron and planetary scientist Kevin Hand see a large gelatinous animal, Hand exclaims, "Look at that thing! That is absolutely unreal!" Cameron says, "Look at this thing. Look at this thing; It's just incredible. . . . Beautiful." The film lingers here, letting us watch the entrancing gelatinous animal billowing like a translucent scarf. Their enthusiastic appreciation of the animal is amateurish, recalling accounts in chapter 2 in which deep-sea vent life was seen and even taken—popped into Tupperware and drenched in vodka—by those who were not trained in marine biology. Despite the entrancing creature in this scene, the film's narrative valences and vertical hierarchies subordinate the ocean depths to outer space. The ocean is touted as the perfect practice arena for space explorers; marine biology is condescendingly cast as a good starting point for astrobiology; and the samples from the ocean are just the next best thing for the planetary scientist to examine. (The sublime, shoved through a pinhole.) Figueroa's compelling and informative discussion of symbiosis in *Riftia,* the giant tubeworms in hydrothermal vents, for example, is followed by a subordinating cut to Cameron telling Hand, "The *real* question is can you imagine a colony of these on [Jupiter's moon] Europa?" Whereas the renowned ocean scientist Sylvia Earle and other blue environmentalists use the comparison between space exploration and ocean exploration to argue for more robust funding for ocean research and conservation—the assumption being that something is wrong when we know more about other planets than we do about the depths of our own planet—*Aliens of the Deep* presumes space exploration is the supreme pursuit.

The film concludes with a scene in which Figueroa, within the submersible, reaches out as if to touch through the glass a billowy, faceless creature that morphs into an iconic space alien (Figure 6)—as if the rippling, serene creature were not entrancing enough. This scene, featured on the cover of the DVD, promises haptic intimacy, as the encounter between scientist and marvelous creature reaches its crescendo in a moment of touch. But the glass separates, and the creature is not what it seems. Worlds apart from Helmreich's potent conception of intimacy with the alien, unfortunately, the documentary fails to imagine human relations with and responsibility to deep-sea life as the space alien provides a vertical escape hatch to nowhere. While it is possible to imagine human kinship with radically different species via our shared evolutionary origins and horizontal genetic transfers, in this scene, the human is literally contained as the scientist is encased within the submersible and, more conceptually, as human knowing and being are separated from the spectacle of ocean creatures. We see, we marvel—it's all cool and exciting—but it doesn't matter. Articulating marine depths with distant planets segregates them from histories and ongoing practices of colonialism and capitalism; they are transubstantiated into a kind of hypertranscendence in which the aqueous transmutes into the ethereal. Casting deep-sea animals as space aliens removes them from the domain of human concern, imagining them as pure alterity. It disconnects the ecologies of the depths from the global flows of oceanic ecologies, conveniently ignoring the human communities who are connected to the zones of pelagic plunder and destruction. Pacific Islanders, for example, live with the legacy of nuclear testing, the decimation of small-scale fishing practices, heaps of plastic pollution, and rising waters. Here the deep seas are cast as playgrounds for the Western explorer. Just before the final scene in *Aliens of the Deep,* Cameron says, "Exploration is like a muscle. You have to exercise it to make it stronger." Echoing Teddy Roosevelt's belief that the United States required wilderness for patriotic man making, the ocean is cast as the last wilderness on earth, with one small change, a DEI maneuver that whitewashes colonial legacies: now, anyone, male, female, Black, or white, is welcome to join the expedition, as long as they embrace the spirit of exploration and cast the adventure as an encounter with alterity.

That iconic gesture of Figueroa reaching out to touch an alien resonates with Donna J. Haraway's brilliant analysis in *Primate Visions* of a Gulf Oil advertisement in *National Geographic* depicting the hands, lightly touching, of Jane Goodall and a chimpanzee. Haraway analyzes the dynamics through which white female primatologists served as mediators for nonhuman primates within the matrices of gender, race, and colonialism, as those constructs are associated with the dualisms of human and animal, culture and nature.[11] Haraway, writing on *National Geographic*'s racial representation of primatology in the 1960s and 1970s, argues that "people of color were constructed as objects of

Figure 6. Scientist Dijanna Figueroa reaches out to touch a salp just before it morphs into a space alien. DVD cover of *Aliens of the Deep,* directed by James Cameron.

knowledge"; they "could not mediate the required touch with nature that could reassure 'man.'"[12] Something different is happening here; the moment of touch between the deep-sea alien and the Black female scientist is not a matter of mediation but of reinscribing alterity. Penelope Ingram in *Imperiled Whiteness* notes that "alien difference is an oft-used metaphor for racial difference, and conservative ideologies of gender and race frequently get reproduced in SF." She explains that ironically, "white SF serves as a precursor to the colorblind fantasies peddled by the postracial, and at the same time, continues to enact and embody latent racial tensions still extant in American life."[13] The fantastical moment of intimacy between Figueroa and the alien reinforces their aligned alterity. By contrast, consider the lively body of music (e.g., Drexciya's *Deep Sea Dwellers*), literature, and theory of the Black Atlantic, which imagines Afrofuturist oceans and cross-species communication, alliances, and material intimacies. Alexis Pauline Gumbs, in *Undrowned: Black Feminist Lessons from Marine Mammals*, for example, quotes Lucille Clifton's line that the "Atlantic is a sea of bones," asking whether those bones dissolved and then speculating about their cross-species transformation: "Sediment. Filtered ultimately into the baleen of an Atlantic gray whale . . . there is actually digestive truth to the idea that the ancestors we lost in the transatlantic slave trade became whales."[14] Gumbs's palpable sense of elemental, multispecies enmeshment reckons with the violence of the transatlantic slave trade yet still insists on cultivating a sense of deep connection and concern for marine life. Despite the material dissolve of human bones into whales, Gumbs's book values marine mammals as beings with particular practices, characteristics, and challenges who inspire "Black feminist lessons" in cross-species alliances and networks of concern and care. Chapter 4 will discuss Gumbs further, but here, we should conclude this analysis of Cameron's imaginaries by stating that *Aliens of the Deep* offers deep-sea life as an aesthetic spectacle, but it fails to present abyssal life as worthy of intimacy or concern.

While Cameron imaginatively transforms deep-sea life into space aliens, Craig Venter transforms oceanic biological matter into codes, endorsing a hierarchy of abstraction, appropriate for capitalist exploits and genetic engineering. Venter, an American biotechnologist and entrepreneur who remains perhaps best known for mapping the human genome, relegates the ocean to a repository of genetic material rather than a site of living creatures, organisms, species, and ecosystems. Helmreich, in his critique of Venter, points out that Venter lacks training as a marine scientist; his method "leaves out essential contexts for understanding microbial sequence data."[15] Moreover, because Venter's vessel was not designed for research, the samples "may be contaminated by effluent from his boat," making them, potentially, "full of shit."[16] Although fecal contamination is not our focus, Helmreich's incisive critique of Venter's conceptual framework is revealing. He explains that when "researchers like Venter describe the objects of environmental genomics as environmental genomes (or even an ocean genome), they slip from a description of their method into a claim that their sample represents a bounded entity in the real world," which makes various things, such as fishes and seaweed but even ocean currents, "vanish."[17] Venter's genetic prospecting also makes life vanish by transformation into abstract code. Featured on the cover of *Wired* magazine with the headline, "Fantastic Voyage: The Epic Quest to Collect the DNA of Everything on the Planet—And Redefine Life as We Know It,"[18] Venter's ambitions are lauded with an inset page, "How to Hunt Microbes," imbuing the search for miniscule creatures with the colonialist masculinity of the great white hunter. This narrative of heroic individualism, scientific progress, and mastery of nonhuman nature reiterates the obsession with genetics, in which life—living creatures, species, and interacting ecosystems—is reduced to codes. Genes—imagined as discrete, mechanistic, agential entities, as Evelyn Fox Keller, Donna Haraway, and others have argued—have become invested with the power of life itself.[19] Venter processes his samples by "shotgunning," or breaking "the long strands of DNA into millions of fragments."[20] The fragments are robotically reassembled by sequencing machines that put the bits back together. Despite the transmogrifications of the microbes, there seems to be no worry that anything is lost, misread, or radically altered in this transformation from elemental life to data. A cavalier epistemological arrogance pervades these exploits. There is no sense of "mangling" in Andrew Pickering's terms,[21] nor any recognition of the entanglement of technologies, procedures, economic incentives, and data. Code seems to float free of messy materiality, intact and uncompromised. As the narrator in the Discovery Channel film about Venter's project, *Cracking the Ocean Code,* announces: "From life to disk": "What was once faint biochemical information of a living creature is now digital code ready for computer processing." The film concludes triumphantly: "And from the sea, a new epoch for all life on earth has begun."[22] Less enraptured, the materiality of the muck from which magical codes are liberated makes a momentary appearance in what looks like gooey pancake batter plopped on a technological device. James Shreeve, the author of the *Wired* article, questions Venter's achievement: "Venter may indeed have 'collected' 100,0000 new species and tens of millions of new genes. Does he or anyone else, possess the conceptual tools needed to pull some great truth out of an ocean of information and vivify it like a bolt of lightning bringing Frankenstein's monster to life?"[23] Shreeve's critique is well taken, yet the "monster" here is already alive, in the form of many (how many?) distinct living creatures that have no need for Frankenstein's bolt of lightning.

Counting: The Census for Marine Life

In the wrenching conclusion to *Before the Law: Humans and Animals in a Biopolitical Frame,* Cary Wolfe argues that justice to some forms of life is predicated on the exclusion of others. Considering how those lines are drawn, he asks who is to be included: "*All* of them? How many? Who knows?," insisting that these "are not rhetorical questions." Wolfe asserts that "all cannot be welcomed, nor all at once" because an "affirmative biopolitics" cannot "simply embrace 'life' in all its undifferentiated singularity."[24] The necessity to choose who will be included means that "in the future we will have been wrong."[25] The ethical imperative in this compressed elegy for all those we will undoubtedly have neglected to welcome is fitting for abyssal beings, especially those who, presumably, are being rendered extinct before their existence is known to scientists and publics. As the *New Yorker* cartoon with which this book began suggests, extending environmental concern to the bottom of the sea seems unlikely, which means that deep-sea species, habitats, and ecosystems may be destroyed though utter disregard. We will never know how many ways we will have been wrong. Yet the ultimate impossibility of solid knowledge and harmless action need not defer ethical practices. The Buddhist vow to save all sentient beings despite how innumerable they may be, for example, endures despite its sheer impossibility. The staggering magnitude of the number of creatures that could be addressed within biopolitical or Buddhist frames gestures toward the possibility that pondering sheer numerical scale may not only spark dumbstruck disinterest, denial, or defeat, but could suspend us within the discomforting space of being implicated in ongoing, shifting, global assemblages of harm, a space where it may be possible to resist, reconfigure, or tread more lightly.

There is no end to reckoning with anthropogenic harms, including the unspeakable losses of extinction. Within global capitalism and the sixth mass extinction, practicing an ethics of nonharming entails epistemological quandaries that are more formidable, convoluted, and far-reaching than those of an anthropocentric risk society.[26] This perplexing mapping of the self—as embedded in economic and political systems as well as within personal, ethical practices in relation to a multitude of other species—may underscore the ecological imperative to establish accurate baselines of species and populations while also revealing that numerical data may be indispensable but is also terribly insufficient in terms of both accuracy and impact. Reckoning with the sense that accountability for anthropogenic harms may be tenebrous and unattainable may expand the domain of environmental concern, as the recognition of the magnitude of the unknown amplifies the vital need for less harmful economic systems, politics, ethics, and practices. This may seem like an esoteric thought experiment, but it aligns with the precautionary principle, an established environmental guideline designed to prevent environmental harm even when scientific evidence may be lacking. Even as ecological and biological science about marine life is crucial, imagining deep-sea creatures as dodging scientific capture—whether that be in terms of numerical data, photographs, or taxonomic identification—could inspire buoyant, if not roseate, feelings. (Recall Percy the deep-sea squid as he glows beautifully blue, or read Ron Broglio's manifesto for the *Animal Revolution.*[27]) While the slogan of the Census, "Making Ocean Life Count," plays on the double meaning of counting to suggest that numerical captures will result in concern, the extraordinarily ambitious—even absurd—aims of the Census may themselves embody another sort of potency. Indeed, as the chapter on William Beebe's deep-sea descents argues, scientific "failure" may foster more capacious, philosophical, aesthetic, and perhaps ethical relations with deep-sea lifeworlds. Paradoxically, both counting and accentuating the impossibility of counting, by conveying the tenuous, fragmentary, and speculative nature of the attempts to tally, can welcome innumerable and unknown species.

An introduction to the ambitious Census of Marine Life is in order. According to Daniel Cressey, the Census was "perhaps the largest and most expensive programme of marine biology ever."[28] The Census of Marine Life website describes it as a

> 10-year international effort undertaken to assess the diversity (how many different kinds), distribution (where they live), and abundance (how many) of marine life—a task never before attempted on this scale. The Census stimulated the discipline of marine science by tackling these issues globally, and engaging some 2,700 scientists from around the globe, who participated in 540 expeditions and countless hours of land-based research. The scientific results were reported on October 4, 2010.[29]

Jesse Ausubel, an environmental scientist, director of the Program for the Human Environment at Rockefeller University, and program manager for the Alfred P. Sloan Foundation; and Fred Grassle, who researched deep-sea biodiversity, including hydrothermic vent communities, as discussed in chapter 2, together founded the Census. Ausubel explains how Grassle approached him in 1996 while holding a National Research Council report, *Understanding Biodiversity,* saying, "This was a good report, but none of the recommendations are being implemented. There's so much to discover and the oceans are changing very fast. . . . You should do a worldwide census."[30] The Census received $78 million from the Arthur P. Sloan Foundation and another $550 million from other sources. Related projects include the International Barcode of Life, the Encyclopedia of Life, and OBIS (Ocean Biogeographic Information Systems), "a global open-access data and information clearing house on marine biodiversity for science, conservation and sustainable development."[31] After the Census concluded, the wealth of deep-sea visual data was synthesized by SYNDEEP during a two-year period, with the help of artificial intelligence, which led to INDEEP, "a global collaborative scientific network dedicated to the acquisition of data, synthesis of knowledge, and communication of findings on the biology and ecology of our global deep ocean, in order to inform its management and ensure its long term health."[32]

Seeking new species, rather than bioengineering them, hearkens back to earlier historical periods. Indeed, biologist Phillipe Bouchet, in "The Magnitude of Marine Biodiversity," published in 2006, argues that even in the 1950s and 1960s "exploring the world to discover unknown species, describe them, and give them names seemed to be a scientific occupation that had its heyday in the 1850s to 1900s." The "general public and policy makers" both assumed that by the end of the twentieth century, "we must surely know the majority of species."[33] Bouchet explains that twenty-five years ago, "scientists believed that the ca. 1.6 million species they had then inventoried represented maybe 50% of plant and animal species on this planet." More recently, however, new "approaches in sampling insect diversity in rainforests and small macrobenthos in the deep sea have revised this estimate to 1.7–1.8 million described species and 10–100 million species remaining to be discovered."[34] It is astounding to imagine so many millions of species that remain unknown to conventional science, some of which (how many?) hail from the depths. In an earlier essay published in *American Naturalist,* Fred Grassle and Nancy J. Maciolek note that "until the late 1960s, deep-sea species richness was thought to be lower than species richness in shallower environments."[35] They note that the "paucity of quantitative data on deep sea species diversity" led J. W. Nybakken in 1982 to conclude that "the idea of a highly diverse deep sea fauna must remain speculative."[36] Grassle and Maciolek, analyzing macrofauna from ten quantitative samples in the Atlantic, finding 58 percent of the species to be "new to science," conclude with this staggering estimate of deep-sea floor biodiversity: "Even very conservative extrapolations of these data to the enormous surface area of the deep ocean suggest that the number of species inhabiting the deep-sea floor has been greatly underestimated. As more of the deep sea is sampled, the number of species will certainly be greater than 1 million and may exceed 10 million."[37] One million! Ten million! The estimates for deep-sea species diversity is astounding, but things become more confounding when we consider that even if these creatures were to be found, they would hover in a kind of scientific limbo for who knows how long? Grassle and Maciolek note that although they are grateful for their taxonomic experts, the Census and other research have resulted in more samples and specimens than can be identified. Bouchet notes, "At the current rate of new species descriptions, it will thus take 250–1,000 years to complete the inventory of marine biodiversity."[38] Two hundred fifty years! One thousand years! The labor, processes, funding, and networked knowledge production of science become impossibly slow, clunky, and ineffectual when set against the staggering number of biodiverse marine specimens. The aims of the Census—to count the diversity, distribution, and abundance of species in the global oceans—could not be more aspirational. But there is more. Given accelerating anthropogenic extinctions, surely many species will disappear before they have been captured, counted, and taxonomized. Their existence as organisms and potential species will dissolve, without regard. As Naomi Oreskes bewails, the Census "came too late": "Scientists were trying to count something that to a large extent was already gone."[39]

Just as the Census was concluding, biologists Michael A. Rex and Ron J. Etter, in their volume on *Deep-Sea Biodiversity,* note that "there are still no reliable estimates of total deep-sea biodiversity."[40] This lack of numerical certainty morphs into the anticipation of future wonders, transforming the lack of tallies into potential aesthetic regard: "Since most of the deep sea remains unexplored we can hardly guess what other wonders exist there."[41] Indeed, Paul Snelgrove and Craig Smith refer to the deep seas as "a riot of species in an environmental calm."[42] Jamieson, less sanguine, notes in *The Hadal Zone* that the Census did not include the hadal trenches, so "the knowledge gap between the trenches and the rest of the ocean is ever widening."[43] Notwithstanding the lack of attention to the hadal zone, the public-facing material of the Census seems to emphasize the deep seas, which is not surprising, given Grassle's expertise in that area, the comparative lack of biodiversity data in that zone, and, most likely, the way in which the newly discovered deep-sea species can be readily disseminated as compelling images, given their novelty, beauty, weirdness, and sensational lifeworlds. While the concepts of species and biodiversity are vital for conservation, both concepts are, of course, deeply problematic, a matter of competing definitions, provisional paradigms, and conditional cuts. Rafi Youatt begins *Counting Species: Biodiversity in Global Environmental Politics:* "'Biodiversity' is difficult to define, precisely. As an ecological concept it refers to the diversity of species that make up life on earth. But as it circulates in the world, biodiversity is at once a natural fact, a species-extinction event, a scientific field of inquiry, a political referent, a moral discourse, an abstract pattern, and a tool of governance."[44] Given the diversity of referents for biodiversity, as well as Youatt's exclamation that a review article from the 1990s discussed "no less than eighty-five definitions of the term,"[45] it would be not only inappropriate for me to select one definition but also counterproductive to the current I trace in this section, which is the way in which the impossibility of the Census is embraced as, dare we say, an aesthetic response. Recall Melanie Sehgal's conception of Whitehead's "nonhumanist" aesthetic as "a *specific dimension* inherent to all kinds of experience," which "takes into account and puts into practice and motion the reciprocity of feeling, relationality and existence."[46] Shifting from epistemologies of objectification, mastery, and certainty toward a sense of the overwhelming magnitude of deep-sea biodiversity steeps the science with an aesthetic of astonishment. How can nonscientists experience their own relationality to innumerable lives in the abyss? How can such numbers be anything but cryptic as they obscure rather than delimit, disperse rather than define? If it is not too much of a stretch to say the magnitude of the numbers of deep-sea species can be experienced aesthetically, in a mode that provokes the "reciprocity of feeling, relationality and existence," then such a response could be understood as countering orientations toward the world that are reductively definitive and objectifying.

The impossible aims of the Census, to classify and count all the creatures in the sea, ostensibly manifest Enlightenment models of knowledge where abyssal specimens are brought up into the light of objective scientific reason. One would assume that the project of counting and categorizing all ocean life—in the past, present, and future—would epitomize the "god-trick" of "infinite vision."[47] Despite the staggering magnitude of the project, Ausubel et al. note that the Census fostered epistemological humility among its researchers: "As it worked, the Census sought to establish the habits to identify, systematically, the limits to knowledge and to speak candidly about the unknown and unknowable as well as about what we know."[48] The first chapter of the coffee-table book *World Ocean Census* is entitled "The Known, the Unknown, and the Unknowable." Many of the essays point out the extraordinary challenges—in terms of scale, sampling techniques, geographic and aquatic hardships, international cooperation of scientists, and the great void of baseline knowledge—of determining what lives in the ocean. Paul V. R. Snelgrove, a biological oceanographer, includes a section entitled "Abnormal Science" in the foreword to *Discoveries of the Census of Marine Life: Making Ocean Life Count.* He explains that the Census is merely a "metaphor," describing an "impossible task *sensu strictu.*"[49] The chapter "Into the Deep" includes an inset explaining "the top ten reasons why the deep sea is special," which includes that it is "the most pristine marine environment left on earth," that the "rate of discovery" of new species "shows no sign of leveling off," that the number of unknown species in a sample "can greatly outnumber the known," and that "new discoveries don't seem to be slowing down."[50] It is important to note that enthusiasm for newly found species in public-facing media of the Census presumes global or universalized publics while erasing Indigenous and other claims to oceanic regions, such as Epeli Hau'ofa's conception of and claim to Oceania as "Our Sea of Islands."[51] Snelgrove's foreword insists that "the ocean is the ultimate global commons, something that belongs to all of us,"[52] which ignores particular claims, including those emanating from cosmologies and kinship relations, to specific regions of and beings within the ocean; ignores the enormity of harm done by global capitalism and the military, including nuclear testing in the Pacific; and fails to mark the devastating impacts of the collapse of ocean ecologies, rising seas, and pollution on particular regions and groups of people. The public-facing texts of the Census are not attuned to social justice or political concerns. Instead, they present their findings to an unmarked, ostensibly universal, global public, echoing Beebe's "brotherhood" with the fish but not the local people, discussed in chapter 1. As with Beebe, the regard for sea life is segregated from historical, cultural, and political relations with groups of people. Even Beebe's philosophical and aesthetic ramblings remain within the conventional parameters of natural science, which trains its eye on nature as a separate domain.

The creaturely orientation of the Census, along with the sense of the impossibility of its own aims, fosters an epistemology that nonetheless diverges from conventional notions of scientific objectivity. In an epigraph from the chapter "Unravelling the Mystery of New Life-Forms" in *World Ocean Census,* Steven Haddock puts the concept of "new species" *sous rature:* "New species aren't really new, they are just new to us. These creatures have been out there for millions of years and we are just now fortunate enough to find them and have the technology to examine them."[53] Displacing the anthropocentrism of the figuration of the "new species," Haddock positions himself as the fortunate heir to both millions of years of evolution and more recent technological innovations. Moreover, he delights in the beings that elude taxonomic grids. He confesses that despite the "natural human instinct to categorize," his "favorites" are "the critters that we have no clue what to temporarily call them. Such animals are given descriptive pet names like 'Mystery Beast,' 'Weird Ctenophore,' . . . 'Blue Bomber' and so on."[54] Finding an undescribed species is, of course, a paradigmatic moment within narratives of discovery, yet the "Mystery Beast," with its "pet name," fosters playfulness and a sense of intimacy. This personal, subjective, and whimsical account seems incongruous with this massive enterprise that no doubt values and must adhere to versions of scientific objectivity. Recall Loraine J. Daston and Peter Galison's question regarding objectivity: "What if the variability of nature is so great as to swamp the infirm human senses and intellect"?[55] The Census, a staggeringly large, technologically endowed project, nonetheless admits that the variability of the global ocean has already swamped epistemological frames. Moreover, the Census, especially in terms of its technologies and databases, including genetic barcoding, results in distributed agencies that channel epistemological crisis away from nodes of subjectivity. Early in the twentieth century, William Beebe admitted in *Half Mile Down,* "Our intellects will never be equal to exhausting biological reality."[56] Admitting scientific defeat in the face of "exhausting biological reality," Beebe contested the narrow confines of modern science while embracing aesthetic, philosophical, and narrative modes of conveying and understanding scientific capture, imaginatively animating the dead specimen and revitalizing the anesthetizing epistemologies of objectivity. By the end of the twentieth century, however, the stakes have shifted, as "biological reality" is less exhausting than exhausted, with declining populations of particular species, the diminishment of biodiversity (however defined), ravaged ecosystems, and of course extinction. The fears that Daston and Galison diagnose seem dusty and quaint: "All epistemology begins in fear—fear that the world is too labyrinthine to be threaded by human reason; fear that the senses are too feeble and the intellect too frail; fear that memory fades."[57] Given accelerating extinctions, contemporary fears pertain to matters more existential than the formation of scientific objectivity. Indeed, considering the disastrous results of the use of modern Western science for capitalist enterprises, the strength, not frailty, of objectivity (and its objectifying maneuvers) is fearsome. Perhaps the fear of indomitable knowledge as domination flickers within the aesthetics of unknowability that flows through the ruminations on the impossibility of the Census.[58]

While Ausubel as a scientist, environmentalist, and leader of the Census sees the limits to our knowledge of the oceans as unfortunate barriers, he also describes unknowability in a more positive sense, especially in his untitled sixty-seven-line poem that serves as the foreword to a book edited by Alasdair McIntyre, *Life in the World's Oceans: Diversity, Distribution, and Abundance.* The poem begins, "The Census of Marine Life is about the total richness of the sea" and then continues in a Whitmanesque fashion, to catalog—composing in the poetic form most akin to a census—that richness. The anaphora "It is about," a colloquial and hazy phrase, is repeated forty-five times. Regarding abyssal life: "It is about radiolarioa and hydrozoa. / It is about lanternfishes and pearlfishes and roundnose grenadiers. / It is about a black, benthopelagic lobate ctenophore and a large pelagic worm with ten long cephalic tentacles."[59] Regarding types of creatures counted: "It is about 10,000 crabs. / It is about 5,000 to 19,000 unique types of bacteria in each gram of sand." And regarding the inability to enumerate: "It is about shimmering shoals of herring swirling in numbers beyond counting."[60] This book about the "known, unknown, and unknowable" is ultimately a "book of mysteries" about "cryptic species." The poem does not only admit but also seems to relish the impossibility of the Census, to delight in the fact that this enumerative project will be swamped not by infirm human subjectivity but by the unchartable biodiversity of the seas. Cynically, we could read this as a way of warding off criticism of the grand but impossible aims of the Census. More affirmatively, though, we can appreciate this epistemology as one that veers toward an ethical sense of wonder at a magnitude that cannot be mastered. Furthermore, the creaturely aesthetic in this poem catalogs marine life as worthy of poetic elevation; each species, even those that are as yet unnamed, is remarkable in its singularity ("a large pelagic worm with ten long cephalic tentacles") while simultaneously astonishing in their numerical magnitude ("swirling in numbers beyond counting").[61]

The chapter entitled "Diversity of Abyssal Marine Life" from *Life in the World's Oceans* underscores both the current and future lack of knowledge about the evolutionary process of speciation and the tally of current species: "We will probably never know the true number of species in the abyss. The research area is far too large to be sampled adequately considering how heterogenous this habitat turns out to be and how high the percentage is of rare species which have been recorded from just one site."[62] The chapter concludes, "Within the past 150 years, we have learned to look at the abyss through different lenses. The unfathomed depths turned from a mythical place inhabited, at best, by fearsome creatures waiting to attack the unwary seaman, to an integral part of our planet filled with a dazzling variety of life, well adapted to the environment and of unsuspected beauty and grace. There are still many more questions than answers."[63] This sobering scientific corrective tames the mythology of the abyss, cautioning against sensationalism while retaining a recognition of both scientific uncertainty and a modest aesthetic regard for "the dazzling variety of life," which may possess "unsuspected beauty and grace." While many supposedly mythical deep-sea creatures still circulate in popular culture, especially in the form of aliens, a small coffee-table book from the Census elevates and anthropomorphizes marine life, more generally, as citizens. The title implies a shift from scientific captures to a global environmental politics in which marine life has a stake, a voice, and some sort of standing. Nancy Knowlton begins *Citizens of the Sea: Wondrous Creatures from the Census of Marine Life* by stating that by the end of the Census, "most ocean organisms remain nameless and their numbers unknown." She insists, however, that this "is not an admission of failure."[64] Whether or not particular marine species have been identified and counted, Knowlton insists they deserve recognition and inclusion: "The sea today is in trouble. Its citizens have no vote in any national or international body, but they are suffering and need to be heard."[65] Knowlton echoes Jeremy Bentham's argument that we should not be asking whether animals can reason or speak, but whether they can suffer. This introductory position is complemented by the visual language of the volume, which emphasizes the liveliness of sea life. *Citizens of the Sea,* a small book overflowing with a hodgepodge of photos, veers away from a cold, objectifying stance and toward a "riot of species."

The cover dramatizes both scientific capture and artistic stylization with its grid, which frames the photos of ten different creatures (Plate 5). The grid signals the overlay of scientific captures onto the ocean, arresting the liveliness of its inhabitants. Yet the subtitle, "Wondrous Creatures," promises awe and amazement that overflows the grid itself, just as some of the creatures cannot be contained within one frame, either on the cover or throughout the volume. The visual agency of the animated specimens performs a kind of resistance to the imposition of the scientific grids, enticing us with the energy, dynamism, liveliness, and magnificence the creatures convey. Moreover, the overwhelming scale shifting of the photographs in this deceptively small volume—some distant shots, some midrange, and some so close up as to be jolting and disorienting—could embody what Zachary Horton calls "the radically nonhuman dynamics and potentials of scale." Horton argues that engaging with what he terms "scalar alterity" implies "a dismantling of the edifices of humanism itself."[66] As the Census calculates the "citizens of the sea," the disorienting effects of the sheer magnitude of the number of beings in the deep—known, unknown, likely to be known, never to have been known—captivates publics by immersing us in the fabulously speculative pleasures of envisioning multitudinous life in the abyss.

Framing: Designing a Reflexive Creaturely Aesthetic

Granted, numerical tallies and estimations—abstract rather than sensual; devoid of color, composition, and texture—are not usually the stuff of aesthetics. One of the most striking elements of the Census, however, has been its production of visually enticing website galleries and internet videos that feature individual organisms. Jody Berland, in *Virtual Menageries,* asks, "Can the global proliferation of digital animal images inspire humans to transform their relations with other animals and the natural world?"[67] While the fact that the visually gorgeous, intimate film *My Octopus Teacher* was followed by the plan to construct the world's first octopus farm dashes any hope for transforming human relations with other species, wherever they exist, images of deep-sea life are nonetheless the primary, even if diaphanous, vehicle for sparking concern for abyssal life. Often described as weird, bizarre, or surreal, the fact that abyssal species seem to be everywhere is itself deeply strange, as terrestrial, digitally enhanced humans, with a few clicks of the keyboard, enjoy seemingly immediate access to unthinkably remote beings. We may be astonished by the biodiversity of extraordinarily distant realms as we live in an era in which anthropogenic threats to planetary life are ever present; even insects that were once common are disappearing. An eerie dissonance permeates encounters with digital images of abyssal life, as the fabulous discovery of an abundance of heretofore unknown creatures is staged against the diminishing populations of familiar, commonplace terrestrial animals. Of course, such images may circulate as mere spectacle—commodified life, clickbait, the far reaches of the world offered up for our momentary pleasure, the sublime shoved through a pinhole. Indeed, Jesse Ausubel, cofounder of the Census of Marine Life, expresses "delight" that "the yeti crab and others of our animals are now replicated as stuffed animals and are featured on skateboards."[68] Nonetheless, against capitalism's unrestrained commodification and the internet's flattening of global volumetric expanses, the aesthetic images of marine life that the Census of Marine Life presented populate even the deepest regions with dazzling creatures. While visual framing contains its subject, it can aesthetically function, paradoxically, not as a mode of separation but as an invitation for the viewer to regard, esteem, or even imaginatively hover with the featured creature.

The most striking aesthetic captures of deep-sea life—for example, on the official Census of Marine Life website and in Claire Nouvian's stunning collection, *The Deep: Extraordinary Creatures of the Abyss*—are photos of a single specimen, presented in a hyperreal manner against a black background of pelagic waters.[69] These photos feature bright, distinctive colors and a sharp contrast between the animal and its dark environs. The beautiful but blatant images would seem to exemplify Enlightenment epistemologies where objective truth exists, unmangled by perspectives, systems of economics and power, or other entanglements. Environmental and feminist theorists have long been critical of aesthetic captures and commodification. The camera follows the gun; the frame distances, contains, and controls its subject; images are bought, sold, and even turned against the beings they depict. The stunning photos of marine animals would seem to exemplify the subject–object divide with its universal spectator, where objective truth is available to all, free of perspectives, systems of power, or other entanglements. In some sense, the taxonomic and enumerating aims of the Census assume this epistemological stance, as scientists discover and portray each organism in a seemingly direct, unencumbered manner. Enlightenment models of knowledge in which science brings something dwelling in the darkness to light seem inherent to the images of heretofore undiscovered species glowing on computer screens. Unabashed realism—indeed, a kind of luminous hyperrealism—reigns. Yet these images may be seen as staged encounters that underscore their own stylization in a way that immerses science, aesthetics, and environmentalism within highly mediated networks of knowing and concern, appropriate for an immersive aesthetic that erodes the divide between reality and representation, nature and culture. The reflexive aesthetic of the Census subtends scientific captures and popular imaginaries; or in a less static sense, the aesthetic flows from the stylized scientific captures to popular imaginaries of the deep.

Website Gallery

In her analysis of biodiversity databases in *Imagining Extinction,* Ursula Heise argues that these global databases emerge "from the conjunction of two tendencies: an encyclopedic, centripetal impulse that reaches back to the Enlightenment and seeks to inventory the entire known world, and the hyperlinked, centrifugal architecture of the Internet, which seeks to approximate a representation of this world through the constant movement between data sites."[70] The website gallery for the Census possesses both those tendencies, at a much smaller scale, as it centripetally includes "New Species" while inviting the viewer to move through the curated photos on the web pages. Although it is in conversation with massive biodiversity databases, it creates a more approachable and overtly aesthetic experience for amateurs. The penultimate tab on the home page of the Census website, "Gallery," pulls down to reveal the options "Image Gallery," "Video Gallery," "Census in the Arts," "Census Perspectives," and "Maps and Visualizations." "Census in the Arts" lists films, exhibits, paintings, sculpture, music, and even postage stamps. The "Image Gallery" includes tabs for "Iconic Images," "New Species," and "Other Images."[71] The conceit of the gallery highlights the aesthetic framing of each creature, presented as a cherished work of art. The digital galleries also hearken back to cabinets of curiosity or wonder rooms, in which the aesthetic value of an object is inseparable from its status as a taxonomic specimen. The Census's gallery transforms an elite venue for exhibiting treasures into a more accessible digital space. While the data sets available at OBIS and the DNA barcoding practices elude the public, the gallery (when it existed) welcomed all viewers with an internet connection. The Census gallery manifests a taxonomic aesthetic, identifying (or failing to identify) a species by showing one representative specimen, but it also does something more.

Despite the scale of the global oceans, the volumetric expanses of the depths, and the differing sizes of the marine life presented, the gallery stages individual creatures in a way that makes them seem not only readily apparent, as they are distinct from their surroundings, but immediate, vibrantly at hand. As Melody Jue perceptively observes: "Species databases involve vast transformations of scale and materiality, resizing the ocean into representational models calibrated to human proportions and sensory dispositions."[72] Jue sees the Census home page as visually similar to a "series of nested aquariums, with differently sized boxes featuring sea creature photographs, or themed links using the icon of a sea creature."[73] Jue's reading of the nested aquariums is apt, but I see the page as a mix between a digital cabinet of curiosity and a portrait gallery, a space where the aesthetic regard for the specimens flows into an ethicoaesthetic regard, an environmentalist concern for each of the organisms as a person. The aesthetic of the portrait obscures ecological embeddedness and interrelations, but it fosters a sense of encounter with viewers, who require no prior knowledge to marvel at the beauty or novelty of each particular being. The attention to aesthetics entices viewers to linger with the creatures, developing attachments. In case the photos are not themselves alluring enough, the captions often emphasize the beauty of the creatures; for example, particularly "attractive anemones" feature "beautiful stinging tentacles" that "vary from a vibrant purple to a creamy brown."[74]

Framing the images transfers an immediate visual sense of value to the creatures included on the site—and more speculatively to the ones not yet enframed. Biopolitically and philosophically speaking, framing is not frivolous. Wolfe explains, "As Derrida, Agamben, and others have reminded us, those who fall outside the frame, because they are marked by differences of race, or species, or gender, or religion, or nationality are always threatened with 'a non-criminal putting to death.' "[75] More broadly, framing "decides what we recognize and what we don't, what counts and what doesn't; and it also determines the consequences of falling outside the frame."[76] In that sense, the visual rhetoric of framing on the Census site performs the inclusion of countless but irreplaceable marine animals, encircling them within environmental concern. The frames on the Census gallery are not subtle; indeed, the frames within frames are repetitive and readily apparent, enacting a process of distinguishing and encasing each creature within an appreciative delineation. The "Image Gallery," for example, is bounded by a bright blue square frame against the black. The left third of the screen features a symmetrical, rectangular image gallery, framed inside sturdy white squares, three horizontally and five vertically. As viewers click on one of the images in a column of frames on the left, the image appears, enlarged, on the right. The right two thirds of the screen features an expanse where the photo of the selected animal hovers, posing a dialogue between the orderly framing and arrangement of the creatures on the left and a sense of fortunate discovery, between knowledge practices and the difficult quest to find and identify marine life. The ratio of known to unknown here is encoded in the distribution of space: the orderly catalog or gallery at the left is half the size of the right (one third of the screen versus two thirds). Even though the space on the right often features just one animal, the specimen hovers within an ample background. We can read this space of unknowing as aligned with an ecological ethics that refuses epistemologies of mastery even as we marvel at the rare glimpse of extraordinary creatures. Viewers' participation underscores a reflexive sense of the creature as being photographically captured and enframed, but not contained, as the flat background, which refuses a sense of perspective, invites us to imagine inhuman pelagic expanses at scales we cannot fathom.

The site encourages its viewers to contemplate each individual image not in isolation but as part of a series, tempting us with over a hundred stunning photos. These images could be read, in Bruno Latour's terms, as a "set of instructions to reach another one down the line."[77] Clicking on the arrows below the three columns on the left leads to more columns of creatures. The fact that the photos are highly mediated embeds the animals within networks that have made them visible.[78] The distinctive web pages exhibit the human labor, the technologies, the fabrications, and the cost of the Census itself. Some of the captions for the stunning creatures foreground science as a process. For example, "This bizarre new copepod, *Ceratonotus steiningeri,* was first discovered 5,400 meters deep in the Angola Basin 2006. Within a year it was also collected in the southeastern Atlantic, as well as some 13,000 kilometers away in the central Pacific Ocean. Scientists are puzzled about how this tiny (0.5mm) animal achieved such widespread distribution as they are about how it avoided detection for so long."[79] Moreover, the first image on the "New Species" page informs us that "twelve hundred new species were formally described during 2000–10 by Census scientists, with another 5,000 or so in glass jars awaiting formal description,"[80] asking us to imagine these photos within a larger network of scientific capture that remains unavailable to the public. If truth is to be found "in taking up the task of continuing the flow, of elongating the cascade of mediations one step further,"[81] then the images pull scientists, environmentalists, policymakers, and activists into their currents. Considering these images not as free-floating postmodern simulacra (via Jean Baudrillard) but more in terms of Latour's sense of "circulating reference,"[82] in which something of the actual creature is transferred through the digital image, encourages a mediated sense of encounter. Berland, critiquing digital imaging culture for being coolly cruel, argues, "Cool is ironic, inarticulate, self-centered, cruel, and private. By generating a cool aesthetic, digital imaging culture contributes to a crisis of attachment as well as a dissolution of narrative."[83] The Census, however, crafted media that foster creaturely attachments through aesthetic regard and stylized staging and framing. Staged visual encounters happen without narratives, encouraging us to imaginatively hover with deep-sea life, unmoored by the extraordinary modes of being that their marvelous forms and unfathomable habitats suggest.

Website Videos

The Census website includes several short videos that pose deep-sea animals as aesthetic treasures that are, strangely perhaps, embedded within the production of scientific knowledge. The voice-over in the video entitled "Hard to See Creatures," for example, introduces the beings as artworks: "They're mini masterpieces. Built with microscopic elegance. The small and hard to see life of the sea. Microbes. Larvae. Burrowers. And they've been photographed and catalogued as part of a massive marine collaboration of scientists across the globe."[84] The first image bears an uncanny resemblance to Van Gogh's iconic *Starry Night,* casting marine life as "masterpieces." A more dynamic, stylized video with a similar title, "Hard to See," follows the first. This fleeting one-and-a-half-minute "Creature Video" features newly discovered animals—one specimen against a black background. Either the creatures were videoed as they were moving, or, more likely, given the obvious special effects, they were manipulated to dramatize not only a sense of motion but also a sense of fortunate glimpses—rare sightings of life in the oceanic realm. The shaky, undulating, shimmering images that briefly come into focus emphasize the difficulties of photographic capture within pelagic expanses.[85] Latour's critique of freeze-framing is helpful here. He defines it as "extracting an image out of the flow, becoming fascinated by it, as if it were sufficient, as if all movement had stopped."[86] While the Census gallery would seem to exemplify a taxonomic and/or commodifying freeze-framing that removes the animals from their lifeworlds, the aesthetic of the videos encourages viewers to value the scientific process as a flow of sightings, captures, analyses, and mediations instead of an unveiling of a static reality. More poignantly, the fact that the creatures come into focus so briefly is fitting for these moments of scientific discovery that occur within horizons of accelerating extinctions.

The video entitled "Deep Sea Creatures," about two minutes long, dramatizes scientific processes more comprehensively. It dramatically juxtaposes two different styles—the colorful creature floating against the black background, and the sudden imposition of graphics—that name the creature; measure the creature; list its family, phylum, and class; or point out features like the light organs on the jewel squid, or the "extendable lower jaw" and "red-light sensitive eyes" of the loosejaw.[87] Icons of measuring sticks and other graphics are briefly pasted onto the moving image. Oddly, the otherwise hypnotic combination of the piano music and the floating, glowing, slowly turning creatures is disrupted by a flash of sound and image, signifying a static interference—the flash of photographic capture, the typing of text, the appearance of scientific graphics. The clashing audio and visual styles of pacific beings versus noisy human technologies dramatize science as practice and process while also underscoring the intrusion of scientists and technologies into the animal's world. While the helmet jelly rotates slowly, a block of text appears and rapidly disappears in the right-hand corner. Was that a tenuous scientific flash of recognition, a conjecture, or a momentary speculation? These postmodern moments break the suspension of disbelief, reminding viewers that the images are not a window or portal into the depths of the seas but highly mediated productions, the fruits of science, technology, art, and design. The videos do not provide the illusion of innocent knowledge but instead invite the public to the scene of scientific discovery, identification, and knowledge production.

Similarly, Jamieson's *Hadal Zone* begins part 3 by explaining the organization of that section: "In an ideal world, the taxa should be split into easily navigable sections in taxonomic order, but the hadal community is not as well known as many other environments and there are several taxa on which a disproportionate amount of information is available. Therefore, the next four chapters are divided into sections that cover the type of organisms found using different sampling methods."[88] The chapters are thus divided into the types of organisms found in "sediment cores," "trawls and sledges," "baited trap," and "baited camera observations or plankton net halls." While Jamieson would like to organize the sections according to the taxa, the structure he uses, which foregrounds the process of scientific capture, dodges conventional epistemologies that distance the object of knowledge from the production of knowledge. Steven Shapin and Simon Schaffer, in *Leviathan and the Air-Pump,* note "matters of fact are held to be the passive result of holding a mirror up to reality." They explain that in "common speech, as in the philosophy of science, the solidity and permanence of matters of fact reside in the absence of human agency in their coming to be."[89] In an obvious way, it is true that deep-sea species exist whether or not marine biologists capture them. Yet a reflexive, immersed aesthetic can disrupt the commonsensical chasms between reality and representation, and by doing so can demonstrate that without the scientific capture of these species, and without their aesthetic dissemination through popular media, the abyss will remain abyssal, imagined as devoid of life or full of monsters and aliens—imaginaries that fail to spark the political will to protect these zones or, more modestly, to reduce the acceleration of anthropogenic harms. An aesthetic adequate for the

Anthropocene would, in my view, suggest that all human knowledge happens through immersion within long circuits of elemental, technological, economic, scientific, and cultural mediation. Even viewers presented with the fruits of these processes consist of the very stuff of the profoundly altered material world. The "Deep Sea Creatures" video, juxtaposing a creature, barely glimpsed, hovering with peaceful music, then disrupted by a flash of photographic capture, static, and an imposition of scientific labels and measurements, potently dramatizes the disruptions of scientific capture, its tenuous reach into the abyss, and its stylized, urgent bid for environmental regard.

Claire Nouvian and the Abyss at Hand

If I had to choose just one book to place on the coffee table inside the emblematic cartoon that began this study, it would be Claire Nouvian's *The Deep: Extraordinary Creatures of the Abyss* (2007) (Plate 6). Nothing could be more bourgeois—overtly signaling taste, education, and culture—than an expensive, oversize, cumbersome, and decorative book set out for guests to appreciate. Nonetheless, Nouvian's *The Deep* is fabulous. Although *The Deep* was not part of the Census of Marine Life, the Census website notes that Nouvian, a film director, "worked alongside Census scientists studying the continental margins to capture some amazing photographs."[90] Nouvian's preface tells how in 2001, after watching "a stunningly beautiful film" at the Monterey Bay Aquarium, her life "changed direction" as she became fascinated by creatures she thought could not possibly be real.

> I was dazzled . . . speechless . . . astounded. . . .
>
> As crazy as it might seem, I had fallen in love at first sight. . . . It was as though a veil had been lifted, revealing unexpected points of view, vaster and more promising. . . . I imagined this colossal volume of water, cloaked in permanent darkness, and I pictured the fantastic creatures that swam there, far from our gazes, the surrealist results of an ever inventive Nature.[91]

When the images from the film faded out, she was disappointed, desiring more. She explains, "I needed a book that would allow me, at my own pace and for hours on end if I wished, to examine these incredible beings that lived secluded in the oceans' darkness. . . . I wanted to know everything about them. . . . I dreamed of a book that would bring together all the most beautiful images captured in that deep sea frontier, inaccessible to the majority of us."[92]

Collaborating with researchers, she creates the book she desired. Lest this sound like a merely personal pursuit, we should note that Nouvian was awarded the Goldman Environmental Prize in 2018 for her deep-sea activism: "A tireless defender of the oceans and marine life, Claire Nouvian led a focused, data-driven advocacy campaign against the destructive fishing practice of deep-sea bottom trawling, successfully pressuring French supermarket giant and fleet owner Intermarché to change its fishing practices. Her coalition of advocates ultimately secured French support for a ban on deep-sea bottom trawling that led to an E.U.-wide ban."[93] A new genus of deep-sea bamboo coral was named after her, *Cladarisis nouvianae,* honoring her work.

Nouvian's rapturous enthusiasm results in a book of equivalent intensity. *The Deep* is a stunning volume, combining brief essays by scientists with breathtaking photographs. Nouvian's efforts involved gathering "well over five thousand photographs," then selecting a mere two hundred for inclusion.[94] The volume is organized into two parts, "Life in the Water Column," comprising more than half of the book, with photos depicting individual creatures that hover in pelagic realms, and "Life at the Bottom," which depicts groups of organisms and different habitats, such as polar depths, seamounts, deepwater coral reefs, hydrothermal vents, methane seeps, whalefalls, and deep trenches. The photos included in the last two fifths of the book are more varied and less stylistically distinctive and will not receive extensive analysis here. Nonetheless, it may be helpful to briefly discuss their significance. The photos in "Life at the Bottom," depicting groups of organisms in various habitats, are more revealing in terms of their ecological information, visually conveying a basic sense of relations between different taxa as well as between species and their environments. They also vividly depict radically dissimilar deep environments, from polar regions to vents and seamounts. This is no small matter because according to Michael A. Rex and Ron J. Etter in *Deep-Sea Biodiversity: Pattern and Scale,* "Deep-sea ecology has not been incorporated into mainstream ecology, which is based on terrestrial freshwater and marine coastal systems. One can scarcely find the term 'deep sea' in the indices of ecology textbooks and major reference works."[95] They voice a now-common refrain: that "since the deep sea remains unexplored, we can hardly guess what other wonders exist there."[96] Rex and Etter include some full-color photos of particular species in the volume, nine of which are individual organisms against a black background and nine of which include the seafloor and/or other organisms in the photo. Of course, photos of pelagic life would not reveal the seafloor, a seamount, or a vent environment. Still, the photos depicting scenes rather than specimens provide more information about ecological relations, such as a bright blue sea star "perched on manganese nodules," a "deposit feeding" bright yellow holothurian "in a manganese nodule field," and sea pens feeding in a group on blue-green sediments.[97] Tony Koslows's *The Silent Deep: The Discovery, Ecology, and Conservation of the Deep Sea* (2007) includes drawings, black-and-white photos, and sixteen color plates, most of which show abyssal species as they crawl or swim or hover within their habitats, sometimes in relation with other species. The plates capture blue or black pelagic waters; sandy or rocky seafloors; abyssal plains featuring an urchin's furrowed track, "brittle stars on a seafloor pockmarked with hundreds of worm tubes"; tubeworms on a smoking hydrothermal vent; "scavenging hagfish" and "red and white bacterial mats" living on the bones of a whale fall; "vestimentiferan tubeworms and the remains of lucinid clams" at a cold seep; swarming vent shrimp around a vent; an "orange roughy swimming over the corals on a seamount off New Zealand," a "benthic invertebrate community on a seamount," a gorgeous photo of a "soft coral and sponge garden" with delicate, intricate, and vibrant corals and sponges living together intertwined. Moreover, the final plate issues a direct environmental warning, contrasting a lively, complex scene of life on a "pristine seamount" with the bleak, barren seafloor of a "heavily fished seamount."[98]

The distinctive photographs that Nouvian assembles in the first three fifths of the volume do not overtly signal environmental concern; nor do they provide rich ecological information,[99] other than the fact that the animals are pelagic, inhabiting the open expanse of the water column, deep enough for utter darkness, devoid of distinguishing geographic features. The entire pelagic zone (lit and unlit) is the largest biome on the planet, with a volume of 330 million cubic miles. The incredible volumetric expanse makes the sense of an intimate encounter with a deep-sea pelagic animal all the more thrilling. Nouvian's artistic crafting does not frame the creature, as the Census gallery did; instead, the black expanse extends to the edge of each page, with no border. The animal is not framed, but it is dramatically foregrounded, creating the illusion of an intimate encounter between the person holding the book and the distinctive being at hand. Nouvian's selection of the photographs and her crafting of the size of the book, along with the composition, scale, and solid background of the images, all contribute to the impact of the volume. The large size of the book, 12 inches by 10.5 inches, allows most of the creatures in the first section to be about the size of one or two human hands, thus enabling each animal to be an appreciable presence for human viewers while leaving enough black space to signify pelagic depths and breadths. The sense of space is paradoxical, however, as the background feels flatly impenetrable and unrevealing; we are denied a rudimentary map and a grounding sense of perspective. Without anything to signify a sense of perspective, the flat blackness seems like a backdrop for a portrait, serving to distinguish the "person" from the background. More speculatively, however, the background signifies a three-dimensional aquatic biome that extends in all directions. The handy vertical scale on the jacket flap, from 0 to 9,000 (meters, I assume), invites us to align an unobtrusive mark on the edge of each photo with the scale on the flap so as to arrive at a number signifying the maximum depth at which each animal was observed. This encourages curious viewers to envision a vertical map of the distribution of the individual animals portrayed—a speculative practice that disorients with its abstraction and lack of grounding as well as its disanthropocentric scale shifting, triangulating relations among viewer, animal, and colossal oceanic space. The vertical mapping of deep-sea life that William Beebe attempted in the 1930s, discussed in chapter 1, comes to life in the early twenty-first century.

The photos included in "Life on the Bottom" do not have the same disorienting effect, as viewers can comprehend the scene as a place, with a ground, a mountain, a trench, or a sort of forest or jungle of, say, tubeworms at the vents. In *Surface Encounters: Thinking with Animals and Art,* Ron Broglio proposes: "To make thought move and do real work at the horizon of the unthought, representation should create a friction, reciprocity, and exchange between the human symbolic systems of representing and the physical world shared with other creatures—the marks and remarks of various *Umwelts*."[100] The solid black background that Nouvian altered to be consistent across the book[101] dramatically distinguishes each animal, but it also leaves terrestrial viewers groundless, unmoored, hovering in the expanse of the water column. This turns the gaze back to the animal itself while dramatizing that the abyssal being dwells "at the horizon of the unthought." The friction and exchange inspire attention that could develop into connection as the viewer imagines the *Umwelt* of each of these pelagic creatures. Nouvian asserts in the introduction: "Our first duty is to *know* the world down below, and to be motivated by the wondrous biodiversity of our planet; not only the manifest biodiversity of the Blue Planet, illuminated by the sun for all to see, but also that of the hidden zone, the Black Planet, with its concealed beauty enshrouded beneath its tons of dark, impenetrable water."[102]

Not surprisingly, with all this mystery and obscurity, references to outer space and aliens creep in. Some photos show white sparks of bioluminescence in otherwise black backgrounds, looking not unlike a starry sky. Most strikingly, the *Nausithoe rubra* jelly—resembling an iconic UFO and surrounded by conspicuous glowing spheres—is described as being "suspended in space."[103] Unlike *Aliens of the Deep*, however, the volume does not impose a subordinating scale of verticality, despite the depth scale on the book jacket. If anything, it is precisely *this* deep world, *these* particularly astonishing animals that deserve attention, not life-forms on other planets. Indeed, in the preface, Nouvian describes being entranced by the movements and transformations of an octopus until it "drifted away, tranquilly, into the darkness of its own universe." Then there are ellipses, followed by a black space between paragraphs, culminating in the exclamation, "What a universe!"[104] Rather than casting deep-sea life as disconnected from terrestrial humans or as the next best thing to discovering life on other planets, she concludes the preface by centering deep-sea animals and stressing our ecological, planetary interconnections by expressing her hope that the book "will have the strength to captivate us and to remind us that we belong to a vast, living chain that is incredibly beautiful, terribly fragile, and—until proved otherwise—miraculously unique in the universe."[105] This chain, on this planet, somehow connects the viewer to animals that seem worlds away.

Nouvian's exquisite composition and layout of the pelagic photographs intimate that each distinct being deserves contemplative regard. Many photos create the sense of an encounter with the creature, dynamically posed in such a way as to express liveliness and agency. The book begins with a two-page spread depicting two different images of gorgeous, bright pink, unidentified gelatinous comb jellies, flowing into two surprisingly different shapes, while the next two pages introduce the viewer to a frilled shark, "the sole surviving member of a family otherwise known from fossils," with a gorgeous jade eye and an open mouth, which seems to be moving toward the viewer.[106] Setting many animals on the diagonal, or just off the diagonal, animates the images, as does the placement of the animals in fluctuating areas across numerous double-page spreads. The captions emphasize the animals' movements, starting with the first two images after the preface that show *Vampyroteuthis infernalis,* with its mesmerizing azure eye, who is "capable of surprising bursts of speed"; and *Stauroteuthis syrtensis,* the glowing sucker octopus, "nicknamed the 'Dumbo octopuses,'" who have "two main modes of locomotion: they can beat their fins or propel themselves through the water by rhythmically contracting their mantle."[107] The "Dumbo octopus," like an open umbrella, elegantly expands on the diagonal, as its earlike fins curve whimsically, even adorably. While many of the animals float in the middle of the page, some seem to be moving toward the viewer, suggesting the liveliness of the creatures and the possibility of an encounter. Dramatically, the two "wings" of a blue translucent *Llyria,* an unidentified species, seem on their way to embracing the person holding the book.[108] Many animals, with their eyes prominently placed, seem to look back at the viewer. Weirdly, the two "enormous telescopic eyes" of the silvery spookfish, *Winteria telescopa,* which cover nearly half its face, seem to peer at the viewer. The abyss stares back! Adding more drama, the right-hand page of this two-page spread is empty except for the very tip of a fin and barely visible bioluminescent glints.[109] Somehow Nouvian manages to simultaneously stage the animal as significantly sizable and to reserve enough space for the background to signal pelagic, aquatic immensity. The most surreal eyes in the collection belong to *Teuthowenia pellucida,* the googly-eyed glass squid, who follows the spookfish, and who is featured, hauntingly, in a different pose on the cover. In both photos, the googly-eyed squid looks down on the viewer with eyes that emerge from cylindrical appendages—white ocular stalks that look weirdly gooey and floppy.[110] We learn from the caption that this squid is capable of hiding from a predator by filling "its cavity with ink, disappearing into the darkness."[111] Given such capabilities, as well as the fact that the "great ocean depths shelter many more species still waiting to be discovered," we are fortunate to glimpse this translucently bright blue squid.[112]

Nouvian's captions animate her descriptions with vivid, whimsically anthropomorphic comparisons that expose the limits of metaphor, analogy, and simile. Here are a few examples: "This pair of flying buttocks is a new species of worm"; "When seen in motion, one would think it was a spider playing an invisible harp"; "Its upper jaw . . . gives it the appearance of a rugby player who forgot to remove his mouthguard"; "A pram bug wields its prey's remains as a shield."[113] Nouvian's ludic, even outlandish, captions delight but also reflexively underscore their own poesis, inviting viewers to creatively encounter the animal rather than observing it as if a veil had simply been lifted from a predictable, static entity. Nouvian's exhibition, *The Deep,* based on the book, was organized to reach larger audiences than the book had. Roughly two and a half million visitors experienced the exhibit, which toured from 2007 to 2016, in Europe, Africa, Asia, and the Middle East.[114] The exhibit featured large reproductions of images from the book, along with preserved deep-sea specimens, stilled beings staged in arid institutions. The specimens "created the opportunity of a direct contact between the public and the collection of fantastic animals."[115] The website assures us that the "creatures have not been stuffed and so what you see is completely natural—no retouching!" The website includes fifteen images of the process: crates of dead or dying specimens hauled up from the abyss; what looks like a blobfish being affectionately held up by Nouvian's face as if ready for a kiss; the encasement of creatures in resin-filled tanks for display. The resin allows the specimens to be exhibited "as close to their real appearance as possible."[116] Without going down the rabbit hole of recent animal studies scholarship on taxidermy, I would say that I much prefer the gorgeous photographic captures of abyssal animals in *The Deep*—which retain a sense of liveliness—to the ghoulish specimens in this exhibit, despite their three-dimensional, scaly, and otherwise textured, size-appropriate verisimilitude. The carefully composed photos—two-dimensional, stylized, and mediated—seem more animated, whereas the exhibited bodies are unquestionably corpses. While the seemingly unmediated immediacy of creatures included in the coffee-table book could be objectifying, commodifying, and domesticating, the marvelous photographic beings shimmer with elusive liveliness. The flat surface of the photographs provokes what Broglio calls a "surface encounter": "If we could think without an inside and outside, if we could be blunt and idiot enough to think without an abyss between humans and animals, we would arrive at another sort of site and productivity—another sort of thinking."[117] Such an encounter, in which abyssal creatures seem to be marvelously, impossibly at hand, along with images from *The Deep* circulating on the internet[118] could entice, lure, and captivate even those who don't care about the bottom of the ocean. Nouvian, citing her own initial reaction to the film she saw at the Monterey Bay Aquarium, asks: "How is it possible that the Earth bears such marvels and that people don't even know about them? Why doesn't somebody just stop the world, for a minute, or maybe just a *second,* to announce that these creatures exist down here, at the very bottom of our planet?"[119]

Why indeed?

4

Clickbait, Black Void, or Intimate Mediation?

Abyssal Aesthetics after the Census

> We enter an alien world, the Twilight Zone, the sea of eternal doom.
>
> —*Blue Planet II,* "The Deep."

> they were not afraid to slow and evolve their breathing. they were not afraid of their kinship with bottom crawlers who could or could not glow. they were not afraid of being touched by what they could never see, never bring back to the light, never have a witness for.
>
> —Alexis Pauline Gumbs, *M Archive*

In the wake of the conclusion of the Census of Marine Life in 2010, images and tropes of deep-sea life have multiplied across media. While the many weird and wondrous aesthetic captures of deep-sea animals have no doubt sparked broad interest in abyssal life, the dispersal of a deep-sea aesthetic may reduce it to a mere style or even a chromatic palette, extracted from the precarious lives and ecologies of the animals themselves. However, aesthetic captures of deep-sea life can also be articulated with ocean conservation, most urgently in the case of deep-sea mining. Mining could begin in 2025 in the Clarion-Clipperton Zone (CCZ) of the Pacific. This has divided Pacific Island nations. Some governments have agreed to extractive enterprises, while many citizens fight for a moratorium or outright ban. *Blue Peril,* a short but potent video called a "visual investigation," explains legal, political, and cultural contexts as it visually depicts scientific simulations of the processes and potential impacts of the mining. This compelling video is enunciated from a Pacific Islander perspective: "As we speak, our source of life and our identity are threatened by deep sea mining."[1] This site of enunciation contests global visions of ocean conservation even as the video itself has been created with organizations that are not based in the Pacific islands. Much of the video consists of visual modeling of the mining and the movement of toxic plumes, but it also includes gorgeous images of marine animals, populating the abyssal realm with beautiful, intriguing, and heterogenous species.

As the narrator begins, "Imagine waking up one day to a lifeless ocean. What would become of us?," an unnamed bright pink animal that looks like a fireworks display of sparkling yarn appears against a black background that glitters with white dots, sparkling like stars (Plate 7). This shot dramatizes the extraordinary creature gently swaying within waters that seem cosmological—an oceanic universe of life that holds this exquisite, fragile being whose round outline evokes a planet—a singular being, a world within worlds, a life saturated with the living ocean. We might recall Epeli Hau'ofa's declarations in "Our Sea of Islands," which contests the colonialist diminishment of island nations as "small" by ignoring their immense marine environs: "Oceania is vast, Oceania is expanding, Oceania is hospitable and generous, Oceania is humanity rising from the depths of brine and regions of fire deeper still, Oceania is us."[2] *Blue Peril* manifests this cosmological vision, explicitly connecting the lovely benthic communities of colorful and delicate-looking organisms with an "us." When the video voices the question "What would become of us?," it refers directly to Pacific Islanders, who depend on ocean ecologies, but it also reverberates expansively, extending the "us" to the beings on the seafloor and even to life itself.[3] The video contests the idea that the CCZ is devoid of life by using a "photographic survey of 88,000 images of seafloor habitat," some of which progressively fill in the screen, seventy-seven squares in a grid, showing one or more than one animal on the seafloor—a revelation of stunning biodiversity and abundance (Figure 7). With so many squares in this grid, the frames are small, which depicts the creatures as diminutive and hard to see, sparking a desire to approach and protect. The video convincingly populates the supposedly empty, inert "resource" of the seafloor with living creatures. The aligned black frames signify the orderly scientific process of the survey as well as echo the aesthetic framing of the Census of Marine Life website gallery. The most compelling scenes use multiple video screens to animate the proposed mining site. One shot includes twenty different videos of active, moving organisms, the frames misaligned and extending off the screen, dramatizing the liveliness of the animals, the ongoing processes of scientific disclosures, and the existence of more animals, off screen (Plate 8). The narrator states, "For the deep sea life of the CCZ, nodules are breeding and feeding grounds, and provide the only hard surfaces animals can attach to. Most of the species that live on or around nodules are found nowhere else on earth." *Blue Peril* continues to warn viewers that "many unique species and entire ecosystems may become extinct even before we know them." Near the end, the video explains how plumes from mining could render thousands of species—some as yet unknown—extinct. As the narrator explains that the plumes "may harm free-swimming species, including whales, turtles, dolphins, sharks, and tuna," the screen becomes a patchwork of seven smaller videos in which such species swim in and out of the frames, their liveliness exceeding the containment of the small videos within the shot.[4] Such liveliness, such creaturely diversity and abundance, could suddenly, violently, vanish. Cut to the simulation of emphatically red plumes, the waste matter from the proposed mining, ominously vibrating and expanding their impact across the mining map.

Blue Peril ends with Alanna Smith of the Te Ipukarea Society of the Cook Islands stating, "Science is just starting to catch up with what Oceanic people have always known. Our ocean is fluid, with no boundaries, everything is interconnected from the coral reefs to the deep seas. And coastal communities rely on the ocean as our source of life."[5] Reverend James Bhagwan, general secretary of the Pacific Conference of Churches, states, "As oceanic people we have a responsibility to our children and our children's children to preserve our kinship with the living ocean, which is so much a part of who we are."[6] This video is remarkable for its lucid synthesis of the science, its simulations and modeling, its Oceanic perspective of profound interconnection with the living ocean, its design, and its incorporation of videos and photographs of benthic and midwater animals. Science and aesthetics saturate each other as the elaborate design of videos within videos captures the remarkable and distinct biodiversity of the depths, as life overflows

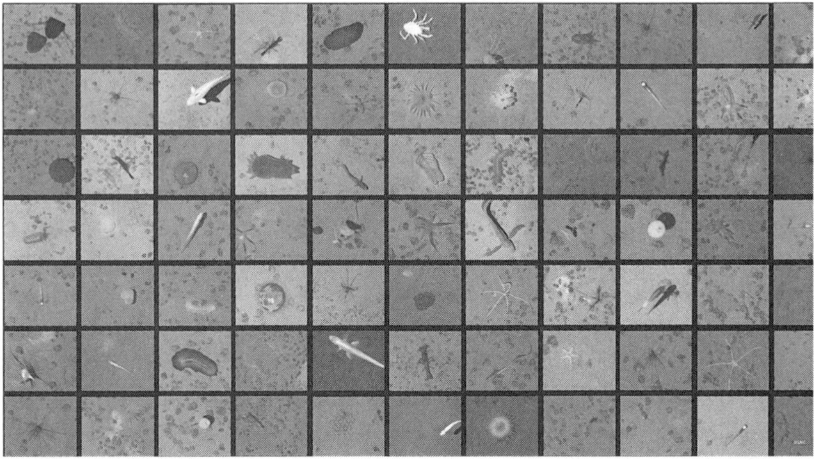

Figure 7. These seventy-seven images are taken from a "photographic survey of 88,000 images of seafloor habitat," demonstrating the abundance of benthic life. Still from *Blue Peril.*

beyond the frames to underscore how this zone, marked for imminent destruction, is interconnected with wider ocean ecologies. Science and technology follow behind the knowledge of Oceanic peoples, whose identity and ethics emerges from "kinship with the living ocean." This consummate model of kinship with marine life is particular to Oceanic peoples and cultures and should be respected as such. While *Blue Peril* underscores the profound connections of kinship and relationality that Oceanic peoples have with marine life—a vital cross-species kinship that would resonate with many other Indigenous peoples—its careful and affecting aesthetic design also opens paths of attachment, alliance, and activism for viewers who are not Oceanic.

Blue Peril voices the urgency of deep-sea conservation at the turn of the second quarter of the twenty-first century. In this fraught moment, when the loss of biodiversity, the acceleration of extinction, and the collapse of ecosystems loom on the horizon, aesthetic modes of encounter with abyssal life that foster mediated intimacies within extinction horizons can perform vital cultural work. While the dominant response to climate change has cast environmentalism as anthropocentric, at least since Bill McKibbens's 350.org in 2008, which concentrated on human communities with little attention to other species,[7] a creaturely aesthetic can evoke disanthropocentric concern for the survival of living beings, species, and ecosystems, by apprehending them as inherently valuable. To call that aesthetic "alluring" reverses the direction of one of the oldest meanings of the English word "lure," used at least since 1386 to refer to the practice of enticing hawks to descend to the human hand.[8] This chapter attends to designs that lure humans to imaginatively descend to the abyss at hand, posting warnings about glittering snares and pernicious depictions of a menacing void. The first decade of the twenty-first century, marked by the Census of Marine Life, brought a multitude of abyssal species to publics in highly aesthetic modes that implicitly argued for the depths as a zone deserving of environmental protection. While this wild amplification of deep-sea discovery and scientific knowledge continues through the first twenty-five years of this century, the scientific and aesthetic captures of deep-sea animals have become iconic enough to be appropriated as merely a novelty, palette, or style. Yet with deep-sea mining on the horizon, the concept of the Anthropocene in the air, and the intensification of many environmental crises, it is inspiring to see the emergence of more explicitly political articulations. Such articulations, across different genres and media, cast a creaturely aesthetic into the sea of public reception as a bid for deep-sea protection. Some of these texts align science with an abyssal aesthetic that counters destruction of the deep-sea biome—a shift that hearkens back to William Beebe's contention that science would not be hurt by aesthetics and emotion. In a time when the scale of anthropogenic extinctions engulfs those who would seek to know within undeniable (albeit murky) networks of culpability, the persistent Western formations of disinterested aesthetic regard and transcendent, disconnected "objectivity" seem not only archaic but also deeply bound up with systematic, even banal, modes of knowing, being, and doing that have caused immense and lasting violence and harm. If, as Lorraine Daston and Peter Galison argue, objectivity fears not only the immensity of the world but also the interiority of the knower,[9] perhaps in the early twenty-first century, the mushrooming scientific and aesthetic captures of the immensity of the world can provoke a sense of mediated immediacy, resulting in more intimate, less disconnected modes of knowing and being that do not externalize "the world" as such.

This chapter, focusing primarily on novels and films, but concluding with a visit to a deep-sea exhibition, traces how a creaturely aesthetic is articulated in ways that encourage curiosity and attachment, but it also analyzes how such an aesthetic is taken up in sensationalistic, distancing, and commodifying modes. Such sensationalism may be saturated with racist discourses that demonize the depths as a black void, antithetical to being, a persistent formulation that Calvin L. Warren analyzes in *Ontological Terror: Blackness, Nihilism, and Emancipation.* The black void insinuates itself within an ominous episode from *Blue Planet II,* to be discussed below. Nnedi Okorafor and Alexis Pauline Gumbs, however, envision the depths as zones of kinship and abundant being, too dynamic and intermeshed for the mechanisms of othering and alterity. The work of Okorafor and Gumbs also presents countervailing and more aesthetically drenched scientific epistemologies, as does the novel and film version of J. M. Ledgard's *Submergence,* and, surprisingly perhaps, the blockbuster creature feature *Meg 2: The Trench.* These works suggest that the conflicts between aesthetics and science that Beebe navigated as modern scientific objectivity was being established have shifted, at least in terms of the representations of deep-sea life in which science and aesthetics are aligned against the forces that plunder marine ecologies. *Submergence* and *Meg 2* also grapple with the concept of the Anthropocene, explicitly or implicitly, collapsing, expanding, and confusing temporal and geographical scales, drenching abyssal encounters with philosophical contemplation as well as campy commotion. While this book has argued for the distinctive role of a creaturely aesthetic for evoking connection, speculation, attachment, and concern for abyssal life, such an aesthetic can become its own undoing as its iconic status, style, or color palette is dispersed, defanged, and exploited in ways that diminish environmental orientations and speculations. The cool weirdness of abyssal creatures, a persistent aesthetic that can inspire posthumanist and disanthropocentric orientations toward creaturely life, can also be severed from species being and cast as a mere novelty, meaningless clickbait, or a diaphanous, gelatinous trifle.

Short-Circuiting Creaturely Aesthetics: Vivid, Creepy, and Cool

In the early twenty-first century, deep-sea aesthetics have become iconic, circulating as stunning, wondrous, or weird clickbait, luring us to bite the hook for another dopamine hit, as nauseating ads for grotesque products pop up all over the screen, rudely tossing abyssal creatures right into the maelstrom of digital capitalism. Even some long-form novels and films exploit the clickbait charisma of abyssal life by severing the potent, affecting aesthetic from the creatures, their lifeworlds, and their precarious environmental moment. This section analyzes how a novel, an animated film, a novelty giftbook, and a video game incorporate abyssal aesthetics without concern for marine lives and ecologies.

The young adult novel *The Girl Who Broke the Sea,* by A. Connors, a British physicist and engineering manager for Google,[10] is set on a deep-sea mining rig in the Clarion-Clipperton Zone, a placid near-future scenario that suggests the mining is a fait accompli. The plot follows Lily, a sixteen-year-old girl who becomes attached to a new species, a bioluminescent euglenoid, with microbial distributed intelligence. Euglenoids, by the way, are protists, neither plant, animal, nor fungi, which usually live in freshwater and are currently of great interest to biotech. As far as I can tell, there are no bioluminescent euglenoids, so the author seems to have infused this sf euglenoid with the captivating aesthetics of bioluminescence as a novum—to enable the plot, but also perhaps to entice readers. Aesthetic encounters lure the protagonist into a relationship with the euglenoid, which begins when she nearly dies descending to the seafloor in an amphipod that malfunctions and leaves her in the darkness. She sees bursts of light, "an infinite parallax of turquoise, going on and on." She continues, "Each dot of light appeared from nowhere and then sparked away, like it was driven by a jet of its own colour."[11] She notices patterns, which suggest the intelligence of this life-form, and receives an invitation to play. She befriends the microbial life via mysterious interspecies communication, and, in a posthumanist philosophical flash, she attempts to explain "what *human* was": "What do you tell a microbial intelligence that evolved in isolation at the bottom of the ocean for billions of years?" Abyssal life here and elsewhere surfaces as the quintessential ground for contemplating the scale of the Anthropocene, as time and space paradoxically seem to collapse into each other and to expand exponentially in baffling ways. Moreover, confronting creatures so different from what humans expect life to be rebounds with the question of what makes humans human within the grand—extraordinarily grand—scheme of things. The novel concludes by saturating abstract aesthetic encounters with the vulnerability of marine life as Lily gazes at "the trembling ocean of light that surrounded us."[12]

This curiosity about "trembling" ocean life that has every reason to fear human incursions, however, is disarticulated from environmental orientations. The sense of wonder lit by the marvels of the abyss is confined to a merely individual experience of one exceptional girl, without the contexts necessary for making sense of the experience within the domains of the ethical or political. The rig is introduced as "the world's first truly sustainable and fully carbon-neutral mining operation," as well as a "unique collaboration of science and industry," one "respectful of local ecology."[13] When three roustabouts overhear this sanguine account of the rig, they exchange looks, but their wariness stems not from the greenwashing but from their fear of the "*things* down there"—"things we haven't seen before."[14] Even as this utopian vision of benign extractivism enabled by collaborations between science and industry is later revealed to be misleading, the not at all dystopian near-future depiction of mining, the novel's light tone, the concentration on the girl's psychological state, the dearth of information about deep-sea mining, and the lack of attention to the diversity of benthic species channel readers away from speculative attachments to abyssal life and concern for ecologies of the Clarion-Clipperton Zone. Note the timing: CCZ mining was to have begun in 2024, just one year after the novel was published. The headlines in popular media, such as this subtitle from the digital site for *Oceanographic* magazine, make the stakes clear: "Over 5,0000 new species discovered in the Clarion-Clipperton Zone, an area threatened by the imminent start of deep sea mining."[15] Given how mining the deep ocean will devastate benthic life and will likely cause as yet unknown harms to already precarious interconnected marine ecosystems, the novel's failure to engage the environmental impacts of the mining is disheartening, to say the least. Finally, when considered within the horizon of global extinctions, we might notice that the protagonist refers to the euglenoid as "the loneliest creature on the planet" as she realizes that the reason she could magically communicate and bond with them is because of their shared loneliness.[16] Reading species isolation as loneliness anthropomorphizes and belittles anthropogenic extinctions. As animal populations dwindle and species disappear, loneliness is not merely an unpleasant emotion but a pervasive biological-social state on the path to extinction.

The now iconic aesthetics of abyssal life may circulate in ways that fail to incite even the most trifling curiosity about the creatures themselves. An odd novelty book by Danielle Kraese, *Deep-Sea Creeps: A Field Guide to Terrible Ex-Boyfriends (As Sea Creatures)* (2024), traffics in the prevalent trope of abyssal life as creepy and weird. Epitomizing the worst stereotypes of heterofeminism, the volume begins with its conceit, the "deep-sea creeps": "Able to thrive under harsh conditions, they can be found in the darkest depths of the ocean, the crustiest corners of your local dive bar, and the most secluded nooks of your D.M.s. If there's one thing they all have in common, it's that you're most likely to encounter them when you've reached rock bottom."[17] The diminutive giftbook is structured with an exquisite painting of a deep-sea animal on one page and a description of a (human) male specimen on the facing page—for example, "Nicolas. *aka* The Ex Who Was Too Cool for Everything," is followed by faux-science field guide descriptions, such as "Notable Behaviors," "Description," "Nesting," "Mating Rituals," and "Range."[18] One learns nothing—nothing—about the animals represented because they are labeled only with their human boy names. Like Nicolas, the book is too cool, but not nearly cool enough. Queer ecologies have no place here as nonhuman life is corralled into heteronormative binaries and perfectly lovely creatures are denigrated by being enlisted to serve as identifying depictions of dysfunctional men. Yet the images themselves, created by mostly unnamed artists for Prince Albert I of Monaco before his death in 1922, are nonetheless seductive; their aesthetic appeal eludes their unfortunate framing, intimating that rock bottom is not without its beauties and pleasures.

Although the giftbook is merely a joke, *Deep-Sea Creeps* epitomizes the short-circuiting of aesthetic encounters as modes of mediated cross-species intimacy. While most of us will never date a cephalopod, we may be seduced by alluring strangers from the depths, whom we encounter in all their gorgeousness, brilliance, and mystery. Such pleasures may occur, of course, as part of a continuum of extractivism, commodification, and even domestication, as the coffee-table book sits obediently on the living room table. Not even the abyss can escape from capitalism and the long reach of Anthropocenic harms. Political and ethical struggles, movements, policies, laws, and practices always take place within the belly of the beast, however—something that the concept of the Anthropocene, with its immense scales and entangled intra- and interactions reverberating through a multitude of sites, ecosystems, and societies, can, depending on its conceptualization, reveal rather than deny. While most depictions of the Anthropocene erase all living species, considering the state of the planet from multispecies perspectives is essential for forging disanthropocentric environmental visions. Thus, it is vital that abyssal aesthetics circulate as creaturely so as not to be severed from species and habitats, and thereby channeled away from modes of attachment and environmental concern. When an abyssal aesthetic is exploited as mere style, it renders the deep sea a frivolous visual pleasure—mere entertainment.

The animated film *Deep Sea (Shen hai)* (2023), which follows the plight of a sad girl who has lost her mother, overwhelms the viewer with a vague, kaleidoscopic abyssal aesthetic that is disconnected from and unconcerned with the animals and ecologies of the deep.[19] The film features a deep-sea restaurant floating on the surface of the ocean, where boorish fish people eat, presumably, deep-sea fish, cooked by wacky walruses (who knows why?). While it would be possible to interpret the restaurant scenes as a critique of the grotesque scale of extractivist consumption of ocean life that extends even to the seafloor, sympathy for the lost, motherless waif drives the plot and drenches the mood. Even as deep-sea species are barely discernible in this fast maelstrom of animation, the supersaturated, vibrant, sharply contrasting, psychedelic swirls of contrasting color upon color upon color suggest the deep-sea aesthetic of glowing digital images that populate screens. This aesthetic—the most notable aspect of the film—is, however, divorced from cross-species attachments; it is merely a style, a sensationalized novelty, taken from the depths and rendered devoid of meaning by ever-churning culture machines. Similarly, the video game *Deep Abyss* (2022), described on the Steam site as "resembling a small painting where you can achieve poetic and artistic sensibility" as you "travel through bizarre landscapes" sells its "vagueness" as interpretive freedom for the players: "some may see flowers while others see the same as waste."[20] In this formulation, individual interpretive freedom squelches articulations that would connect aesthetic encounters with the ethics, politics, and practices of environmental justice as well as ocean conservation.

While chapter 3 argued that the careful framing and design of portraits and image galleries from the Census of Marine Life's website and

Plate 1. With four jars of fanciful specimens in the background, a piscine-looking Beebe bends over to salt and fry a cheerful fish. Miguel Covarrubias, "Professor William Beebe," in No. 2 of the Private Lives of the Great series, *Vanity Fair,* November 1933, 29. Library of Congress Control Number 2016680404. Image courtesy of Marina Elena Rico Covarrubias.

Plate 2. Beebe's giant fishlike head is accompanied by deep-sea fish and mermaids as his body extends from a bathysphere portal. Miguel Covarrubias, "Professor William Beebe and Professor Piccard," Impossible Interviews series, *Vanity Fair,* April 1935, 29. Library of Congress Control Number 2016680252. Image courtesy of Marina Elena Rico Covarrubias.

Plate 3. One of Bostelmann's most exquisite works, in terms of the color, composition, detail, and expressiveness of the fish. The palette and composition are alluring, the fish whimsical. Else Bostelmann, "Saber Toothed Viperfish," WCS Archives, WCS-1039-01-05-1001-R-CR. Copyright Wildlife Conservation Society. Reproduced by permission of the WCS Archives.

Plate 4. One of Else Bostelmann's most comic, cheery, and anthropomorphic works, "Front View of 8 Deep Sea Fish," was also playfully entitled "Big Bad Wolves of an Abyssal Chamber of Horrors." Even the fiercest-looking creatures, with gaping mouths, are depicted as more adorable than monstrous; affection permeates the scene. Else Bostelmann, "Front View of 8 Deep Sea Fish." WCS Archives, WCS-1039-01-05-1033-R-CR, copyright Wildlife Conservation Society. Reproduced by permission of the WCS Archives.

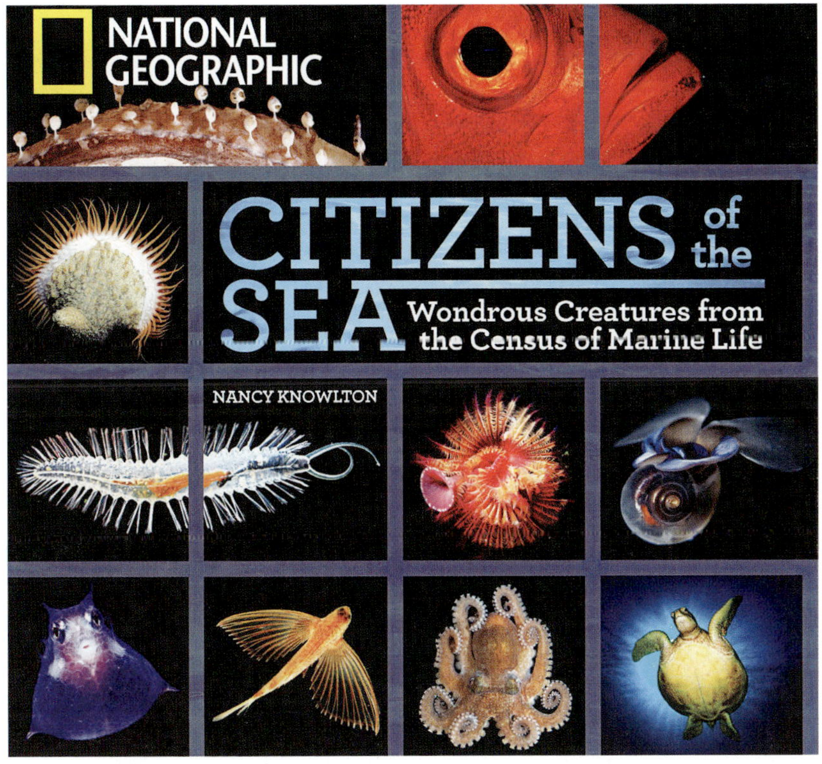

Plate 5. The cover and the overall design of *Citizens of the Sea,* by Nancy Knowlton, echoes the Census of Marine Life internet galleries with their explicit frames and grids, suggesting both scientific capture and aesthetic appreciation.

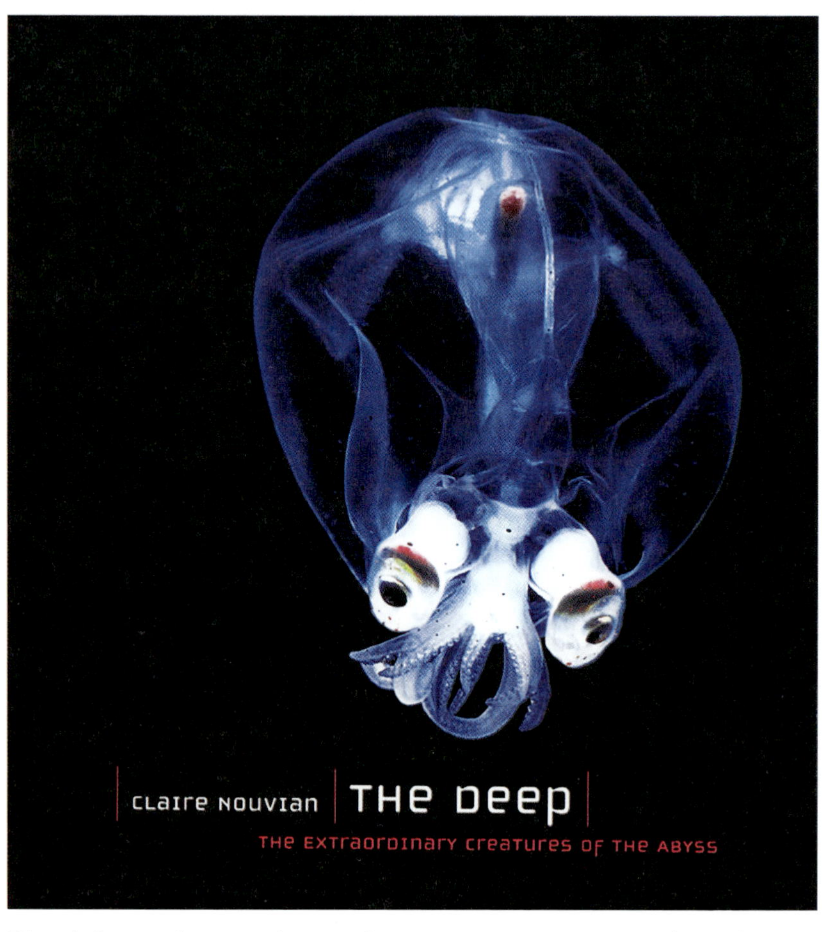

Plate 6. Seemingly gazing down at the viewer, a mesmerizing googly-eyed glass squid graces the cover of Claire Nouvian's dazzling volume, *The Deep: The Extraordinary Creatures of the Abyss.*

Plate 7. An unidentified glowing pink creature, shaped like a planet, sparkling within pelagic expanses that glitter like starry skies, radiating cosmological expanses. Still from the film *Blue Peril.*

Plate 8. Twenty video screens, some exceeding the bounds of the shot itself, highlight the many lively animals who swim into and out of the frames. Still from the film *Blue Peril.*

Plate 9. This octopus is one of the many vivid and extraordinary animals depicted in "The Deep" from *Blue Planet II.* Still from *Blue Planet II,* "The Deep."

Claire Nouvian's book encouraged aesthetic pleasures of marveling, curiosity, and connection, such creaturely aesthetics require care and tending; the route from aesthetic encounter to environmental concern is as fragile and amorphous as gelatinous animals. Disconnecting an aesthetic of the deep sea from the animals themselves obscures the route from highly mediated but seemingly intimate encounters with deep-sea life that spark a sense of appreciation and attachment. Such attachments can not only inspire concern for the particular animal represented but also impel ethical and political commitments and practices directed toward minimizing anthropogenic harms to the abyssal realm, which is always at hand, in terms of quotidian networks of anthropogenic harms. The inaccessibility and sheer distance of the deep would seem to preclude any sense of intimacy with people who live on the surface, yet I hope the oxymoronic phrase "mediated intimacy" captures how the abyss comes to be at hand through long networks of scientific captures, stylized designs, and speculative mappings, which offer forays into creaturely lifeworlds while at the same time grappling with the reach of anthropogenic harms.

Science, Aesthetics, and Mediated Intimacy

While William Beebe's aesthetic embrace of abyssal life in the 1930s was admonished as not being properly scientific, the first quarter of the twenty-first century has seen a strong current that aligns the science and aesthetics of the deep, an alignment that is articulated with environmentalism and against capitalist assaults on the planet. Given the mid- to late twentieth-century mechanisms of extraction, containment, and manufactured ignorance discussed in chapter 2, this formation seems momentous. The fact that such an alignment is dramatically displayed in a blockbuster creature feature such as *Meg 2: The Trench,* discussed at the end of this section, evinces its potency and prevalence. But first, we begin with J. M. Ledgard's 2013 novel, *Submergence,* in which one of the protagonists, a biological mathematician studying the hadal zone, had been a "proponent and a player in the Census of Marine Life."[21] This slim novel poses several interpretive problems, especially in terms of its structure, which juxtaposes the story of James, an English spy held captive by jihadist fighters in Somalia, with that of Danny, the scientist who descends into the depths. The romantic encounters between James and Danny are too fleeting to constitute a romance plot, as scientific and philosophical ruminations expand well beyond personal drama. Their connection, however, poses the question of the relation between deep-sea science and geopolitical conflicts, which could be understood in terms of the legacies of colonialism, and which veer off into radically different domains, sites, experiences, and temporalities—from the long histories of colonial science with its global visions to current incidents of war and terrorism. While such things may seem unrelated, the novel encourages us to speculate across immense scales that overwhelm the tenuous romance plot. Most notably, the novel ponders the time scales of the Anthropocene, leaping both backward and forward from the present, focalizing these meditations through Danny but depersonalizing them, so that they hover beyond an individual consciousness or psychology. Danny ponders a time before submarines and anthropogenic noise, when life was abundant in the Greenland Sea, before the ocean was "being fished out, poisoned and suffering acidification." As she ruminates on a time before the Anthropocene, defining it as "a geological age marked by plastic," the scene positions recent deep-sea science, focalized through this particular scientist, as emanating from within the current moment of environmental devastation.[22] However, the scene also surrounds the novel's present time with longer epochs both before and after the Anthropocene. These are not merely abstract, transcendent mappings, however, because Danny ponders the temporalities of particular life-forms. Danny's reflections swirl together rare microbes that wait a "million years" for "conditions to change so they could become abundant" with queer dimensions of being: "The microbial life of the deep exists in the queerest plane, where worms live in scalding pools and keep fleeces of microbes on their backs that are even more extraordinary than those that live on the timbers of our eyelashes."[23] The creaturely aesthetic revels in the weirdness of life in the deep without alienating it from human experience, appreciating it as "more extraordinary than" the microbes on our eyelashes, but with the comparison itself making the "timbers" of our eyelashes strange—and, more crucially, casting readers' own bodies within breathtaking expanses of time and space, with abyssal worms and human eyes linked through evolutionary webs of kinship and relations with microbes and trees. This is a scientifically mediated mode of intimacy with other beings that exist in their own times and places, astoundingly different, yet, in evolutionary terms, interconnected, as in Charles Darwin's vision of the "tangled bank," with its "grandeur," or Darwin's depiction of the human body as queerly inhabited by relics from evolutionary ancestors.[24]

As Danny and the rest of the scientific team descend, she reflects on how in the abyss, "everything spoke in light," noting that the "puniest fish had the brightest lanterns."[25] Aesthetic marvels abound, such as the "fish who wore a cape of silver chain mail to reflect light" and the squid who "appeared to be encrusted with emeralds and amethysts," with a "massive sapphire eye."[26] As they descend to the vents, Danny perceives the vents as being "in the style of Gaudi," appreciating how the structures provide habitats for the dancing amphipods and swaying tubeworms.[27] Such aesthetic reveries intermingle light, color, fashion, jewels, and architecture, culminating in a musical, metaphorical question: "Did the abyss sing of itself?"[28] As with Rachel L. Carson in *Under the Sea Wind,* the scientist's aesthetic appreciation dilates into a multispecies meditation on how the animals themselves experience light, textures, colors, and sounds. Asking whether the abyss sings of itself encourages us to hover in speculative practices that ponder how abyssal animals experience their own worlds, underscoring that they are not aesthetic objects but living beings with their own perceptions and orientations. The self-referential shorthand, the abyss singing of itself, shifts from the human recognition of beauty to a disanthropocentric creaturely aesthetic. Such an aesthetic, as we have seen throughout this book, most notably perhaps with Else Bostelmann's surrealism, has long been articulated with the depths. It is fitting that *Submergence* concludes with an epitaph from Horace: "Plunge it in deep water: it comes up more beautiful."[29]

While the novel encourages meditations on deep-sea life, evolutionary time scales, and the beauty of the abyss, the 2017 film *Submergence,* directed by Wim Wenders, relays those themes through the embodied presence of Danny, played by Alicia Vikander. The film begins by visually plunging viewers into cerulean waters, lit by the sun above, populated with constellations of gelatinous life. Danny's face, peering out of the helmet of a dive suit, looks enthralled by a benthic scene, suggesting an aesthetic response to the seafloor, even as a recording of scientific descriptions of the vent systems, delivering detached information in official tones, competes with the impassioned music for the audience's auditory attention. Such auditory layering intimates an intermingling of scientific information with an emotional, aesthetic response, as seen on Danny's face and as heard through the music. Danny is not hearing this music herself; it is the film's soundtrack. Instead, the music conveys Danny's aesthetic response to the science, luring the viewer into an emotional connection with scientifically mediated information. The film fools us, however, when this scene concludes by revealing that the seafloor Danny is walking on is actually a set. She is within a lab, testing the suit. Yet the emotional and aesthetic potency of the scene is nonetheless real, even as it is mediated through the scientific information and the fabricated set—intensely real for the character whose experience of this simulation is drenched in attachments to and speculations about life in the deep, as well as for the viewers, who feel the allure of the abyss through the focus on Danny's expressive face as well as through the affecting musical overlay. An aesthetic, emotional appreciation of abyssal life—the awe and wonder conveyed by the mediations of scientific information and visual captures—can inspire speculative engagements as well as affective connections and attachments. The film's inclusion of a set that viewers confuse with reality could be seen as a postmodern mise en abyme, a representational free fall into simulacra. In this twenty-first-century moment, however, I think the scene dramatizes how science and art are practices immersed within the world they attempt to understand, suggesting that the intermingling of science and aesthetics is necessary for Anthropocenic speculations across time and space. Jean Baudrillard's postmodern conception of simulacra, which replaces reality with representation, can be understood, avant la lettre, as diagnosing another vector of the Anthropocene, as cultural production overwhelms and obscures the reality of nature.[30] The twenty-first century recognition of the enormity of environmental devastation, however, suggests the necessity for more immersive and mediated modes of knowing that can transmit, contest, and design (post)natural realities.

Whereas the novel portrays Danny's musings about the deep sea in a rather depersonalized voice, more like an essay than a stream of consciousness, the film cannot resist eroticizing her, most blatantly by showing her in a swimsuit at the beach. Channeling aesthetic encounters with abyssal life through the embodied presence of the alluring female scientist, the film tilts toward an ecoeroticism, connecting terrestrial humans with the depths. When Danny asks James to close his eyes as she seductively intones that the hadopelagic realm is what interests her, for example, she simultaneously arouses his interest in "this other world in our world" and in herself. As she awaits her opportunity to descend to the bottom of the ocean in a submersible hoping to "make a case for [her] theory of life down there," their sexual encounters become consonant with the astonishing reach of life itself. Near the end, after the submersible team survives systems failure and successfully obtains samples, the film concludes with Danny looking out the submersible's portal. This scene occurs just after James has plunged into the ocean to escape the bombing of the jihadists by helicopter, with a close-up of his mouth screaming into the sea. Cut to Danny, her serious face peering out of a portal, a tiny circle in an expanse of blue-black that becomes brighter as she ascends. She is contained within the submersible, acting as the viewer and knower while also being herself framed and exhibited to the film's audience. The sequence cuts back to James, who is floating—dead or barely conscious—in an eerie bright yellowish light. Danny then appears to him in this light, which would seem to conclude the film in a conventionally romantic manner. Yet his death links him to her mythical obsession—a postextinction narrative in which people "will take our place among the teeming hordes," merging with abyssal life in a "submergence." This narrative responds to the existential threats of the Anthropocene by counterposing geological and evolutionary temporalities that exceed and envelop the human. The figure of the scientist encountering abyssal life in emotional, aesthetic, and even mythical modes echoes Beebe's contention that biological science would not be harmed by being receptive to the aesthetic and emotional. Beebe, in the 1930s, however, was not grappling with the magnitude of anthropogenic extinctions and the threat of ecological collapse. The scientist in *Submergence* seems to manage the weight of these concerns by pondering multiple temporalities as well as the persistence of life beyond the human.

I now shift from this contemplative, moody film to the action-packed, campy, and politically savvy blockbuster, *Meg 2: The Trench*

(2023), which follows a scientific team that must battle a deep-sea mining corporation as they contend with megalodons—gigantic sharks, living fossils from the depths.[31] This conflict aptly mirrors the worry that scientific exploration of the abyssal zones will, even if unintentionally, expand the range of capitalist exploitation. In this case, Mana, the research group, is used as a cover for the deep-sea mining of rare earth minerals in the Mariana Trench. The secret mining operation takes advantage of Mana's research site, but without informing the scientists. The revenge-of-nature motif, common in creature features, is appropriate for an era of anthropogenic extinctions. In this film, it is even more so, as the staggeringly immense expanse of anthropogenic destruction corresponds to the gigantic size of the megalodons as well as the mammoth time scales of their existence. The megs have managed to survive from the Cretaceous period to the present, but now, with their abyssal home presumably disrupted by the mining operations, they are chasing and chomping as many humans as they can. Unlike the mysterious whatsits of John Wyndham's *The Kraken Wakes,* the megs are hypervisible—huge, dull-gray sharks, unappealing but impossible to ignore. These megamegamegafauna are not charismatic. They do not need be adorable, however, because they do not need saving. The deep sea has long been imagined as a safe haven for creatures from other geological epochs, motivating Wyville Thomson and William Carpenter in the late nineteenth century, for example, to undertake deep-sea dredging projects that "pass into the modern era of oceanography as Big Science."[32] At the turn of the twenty-first century, the desire to discover fantastically enduring species hiding out in the abyss can be understood as an expression of extinction anxiety because the concept of living fossils can alleviate the guilt and fear that global capitalism has committed large-scale ecocide. Living fossils also radiate the hope that somehow life itself will reemerge from the depths, vitality and fabulousness intact.

In their introduction to the genre of the creature feature, Bridgitte Barclay and Christy Tidwell note its characteristic mix of "pleasure, fear, horror, and sf," arguing that the "absurd form and content" of this genre helps viewers "defang" the "absurdities outside the film."[33] *Meg 2* fits this bill, as many of the scenes in which the megs devour people are hilarious—*chomp!* Yet the bond between one of the megs and one of the scientists also poignantly suggests that cross-species trust, affection, and goodwill are possible even when utterly implausible, even when traversing geological epochs. But what is being defanged here? Perhaps the conceit of the Anthropocene itself—the self-aggrandizing idea that the human has successfully mastered the planet. The megs crash the party, bursting into the Anthropocene from a distant geological epoch, raging that nature is not dead yet, thank you very much. The abyss bites back! But it is a fleeting, aesthetically radiant scene that we should consider here. While most of the film depicts chasing, fighting, escaping, and blowing things up, when the scientific team cruises through the deep-sea vents, we enter a tranquil fairyland. As the main cohort of scientists and other characters enter "the unknown" area, walking along the trench seafloor, at twenty-five thousand feet down in special dive suits, a wondrous scene unfolds. They exclaim, "Amazing!" "It's incredible!" "There are new species all around us!" The daughter, in awe, remarks that it is "so beautiful." One shot sets the team in the far distance, their small white flashlights barely discernible, with glowing aqua and purple bioluminescent animals populating the bottom half of the screen in the foreground, an expanse of blackness above. The darkness extends into the unknown while the beauty below captivates. One shot places what looks like blue coral on the upper right and lower left of the frame, designing on the diagonal, filling out the frame with waving and glowing deep purple stalks, dotted with a constellation of bright green lights. The shot could not be more artistically composed, or more fleeting. A snake suddenly shoots out of a gorgeous bioluminescent creature, attacking one of the scientists and dissolving the fairyland into comic horror. Another team member warns everyone to not touch anything—an enduring principle of conservation, appropriate for this inhuman terrain. Ephemeral and discordant with the rest of the monster movie, this fairyland scene nonetheless demonstrates how a vibrant aesthetic has come to signify the inherent value of abyssal life, its resplendent displays implicitly arguing for its protection. Despite the loud, action-packed creature feature mash-up, we hear, or more appropriately see, the lament of deep-sea biology and conservation: capitalist rapacity, in its protean forms, could destroy deep-sea life before it can be scientifically or aesthetically captured, hurling who knows how many species into extinction.

From Ontological Terror to African Futurism and Speculative Black Marine Biology

While the horror in *Meg 2: The Trench* is wonderfully campy, more sinister renditions of the abyss await us. Many horror films and novels have been set in the deep sea and populated with monsters. Recent novels in this vein include Mira Grant's mermaids in the Mariana Trench horror novel *Into the Drowning Deep* and D. T. Neal's tentacular horror novella about a living fossil, *Relict.* What interests me here, however, in this study of science and aesthetics, is how horror creeps into a reputable scientific documentary. Nature writing and documentary films have a long history of sensationalism that could be traced to the nature fakers controversy at the start of the twentieth century. Nonetheless, the highly respected Blue Planet series, resonating with the gravitas of David Attenborough's voice, and seemingly oriented toward fuzzy environmental awareness, if not actual concern, may leave viewers confounded by the episode "The Deep" in *Blue Planet II* (2018). This episode is a marvel in terms of its breathtakingly crisp, hyperreal close-ups of deep-sea life, as well as its disparate settings, which include areas under the ice, deep-sea corals, a brine lake, and vent systems. The visual captures, brought to us by science, technology, and filming artistry, are exquisite. The episode begins with a sense of stunning abundance: "Astonishingly, in the deep sea there is more life than anywhere else on earth."[34] This theme is accompanied by gorgeous, dazzling scenes and high-resolution close-ups of a multitude of abyssal animals (Plate 9). Gasp! Swoon! As we descend two hundred meters down in the Pacific, however, there is foreboding: "We enter an alien world, the twilight zone, the sea of eternal doom." The trope of the alien disconnects the depths from the rest of the planet, while "eternal doom" transforms this zone of abundant life and beauty into something forever dismal and bleak. Why? The "strange creatures" featured here include a squid with one eye looking down and another larger eye looking up, and, "even stranger," "a barrel eye, a fish with a transparent head, filled with jelly, so he can look up through his skull." Contemplating these intriguing, gorgeous, and remarkable animals can provoke a furtive, closeted, creaturely attachment in spite of the narration, tropes, and editing that segregate such desires.

While the animals themselves provoke awe and wonder, framing them as aliens blocks the aesthetic from currents of attachment and environmental concern. Indeed, the conflicts this episode stages are primarily those between deep-sea animals themselves or between the animals and their environments, not between the animals and various anthropogenic incursions, disruptions, and devastations. When Humboldt squid find a shoal of lanternfish, grabbing them with their tentacles and devouring them, the music is suspenseful. Then Attenborough announces, "When there are no more lanternfish to be found, they turn on each other." Dramatically, Humboldt squid not only attack each other, but attack the camera—more than once. As well they should! Their own image is being turned against them. As the next scene moves on to "the world of perpetual black, below," the midnight zone is called "a giant black void, larger than all the world's habitats combined." Since the episode begins with the theme of abundant life, it is jarring to hear this magnificently vast habitat called a "void." The multitude of astonishing close-ups of the truly stunning, exquisite, and diverse deep-sea animals and the many sumptuous scenes, such as the fabulous blue fireworks-like explosions of bioluminescence, ignite curiosity, even awe. This viewer desires more, wanting to linger with these marvelous beings, to hover in their volumetric worlds, and to resist the film's hasty pace. Indeed, such rapidity, which overwhelms the viewer with novelty, feels more like clickbait than a nature documentary. Ultimately, however, the desire to marvel at abyssal life is short-circuited by horror tropes and manipulative music that warns of the alien, black void.

A long history of anti-Blackness appears to have descended to the bottom of the sea. Calvin L. Warren, in *Ontological Terror: Blackness, Nihilism, and Emancipation,* referencing a quote from Heidegger, contends, "Nothing is the essence of science—the void, the abyss, the unruly thing is the repressed ground of scientific inquiry."[35] If science "must repress this nothing to proceed scientifically," then it does so through "embodied projections of this nothing." Black "~~being~~," (in Warren's terms) serves science: "Blackness enables a scientific encounter with the horrors of an entity that is nothing and something at the same time."[36] Warren's analysis tracks with the episode's contradictory themes of abundance and emptiness, beauty and fear, wonder and warning. Such nonsensical contradictions signal that we are swimming against ideologically saturated currents. Warren's arguments also illuminate why shots of the submersible are included. By including the submersible in several scenes, viewers can identify with the safely encapsulated (white) knowers, masters of technological marvels, who shine their bright lights on the darkness. The knowers are unseen, disembodied, "universal," laying claim to all that they survey. While the Census videos self-reflexively overlaid scientific terms and measurements on the animals that the camera caught, in a mediated aesthetic appropriate for the Anthropocene, here, placing the submersible in the shot reinstates conventional epistemologies that separate subject and object, knower and known.

Just as the ominous warnings about the depths as a black void resonate with long histories of anti-Blackness, the continual invocation of the alien as threatening can't help but resonate with racism against Brown people from south of (down below) the U.S. border. Even the abundance of life in the depths—a "problem," perhaps, that the ideologically saturated scenes seek to resolve—could be seen as a threat to whiteness—too much difference, too much diversity, with the "teeming hordes" below. Warren begins *Ontological Terror* by asking, "Can blacks have life?" And if so, what "would such life *mean* within an antiblack world?"[37] Given the pervasive brutalities of racism, I do not think it is overreaching to see "this (non)relation between blackness and Being" that Warren examines as insinuating itself into a documentary about the deep sea. The abyssal zone is, to be sure, perfectly habitable for the animals who live there; the darkness is exactly as it should be. This vital and abundant biome is black, if not Black. While the azoic theory of the mid-nineteenth century, which asserted that the pressure, cold, and darkness of the depths would make life impossible, was discredited long ago, the "(non)relation between blackness and Being," in Warren's terms, lives on. In their essay "Fear and Loathing of the Deep Ocean," marine biologist Alan Jamieson and his colleagues critique this episode for its "enigmatization" of the depths, noting that none of the other episodes emphasize "how much we do not know about a subject to create 'mystery' through contrived enigmatization" for its entertainment value "while leaning on the viewer's subconscious thalassophobia for effect."[38] I agree with this assessment, but I would add that the thalassophobia is drenched with racist horror.

The repetition of horror tropes—within the narration, sound effects, music, and specific images—also short-circuits any environmental aspirations this episode may have. Sure, there is one scene depicting the beauty of deep-sea corals and the lavish life they support, concluding with the sad scene depicting how trawlers have "ransack[ed] the deep," leaving "countless numbers of the reefs that have flourished here for millennia" lying "in ruins." But while the destruction of deep-sea corals is an outrage, this scene is flanked by two others that wildly upstage it, one that demonizes sharks and one that showcases the grotesque suffering of deep-sea eels. Evoking fear and disgust with sharks and snakelike eels, given their already demonized status, is cheap sensationalism. In the shark scene, the ominous music, creepy sound effects, sight of blood, and the statement "each bite releases blood into the current" demonizes the sharks, which is absurd not only because they are vital part of marine ecosystems but also utterly ridiculous when we consider that they are eating an already dead whale that has fallen to the seafloor. They are scavenging, not killing. Moreover, the sharks helpfully expose the whale's bones for other animals to consume. This is a food web, not a massacre. Ecological relations and environmental concerns are crushed by shallow sensationalism. The episode neglects to blame the long history of industrialized whaling, the continuing sonic violence of the military and commercial shipping industries, and the emptying of the seas from industrialized fishing for the reduction in the number of whales that now fall to the seafloor. To what extent has the decimation of cetacean populations diminished benthic and abyssal ecosystems? That question is not raised. Moreover, the scene that follows the destruction of coral is the strangest and most memorable in the entire episode. It begins with methane explosions on the seafloor but then centers on a "brine pool," a weirdly confounding and disorienting sight of a lake at the bottom of the sea (Figure 8). As cutthroat eels search for food, some "venture into the brine," experiencing "toxic shock," which twists, contorts and even knots up their twitching bodies in grotesque and horrific ways. (You can't unsee this.) This is where the film forces viewers to linger. Some eels escape, but for others, "the brine embalms their bodies and the casualties accumulate among the margins." Rather than valuing the bodies as part of an ecosystem that will feed other animals, the depths become a ghoulish graveyard. Shifting rapidly—making a quick escape—from the demonized and grotesque to the origins of life itself, the episode concludes with vents in the Mariana Trench, noting that life on earth may have begun there.

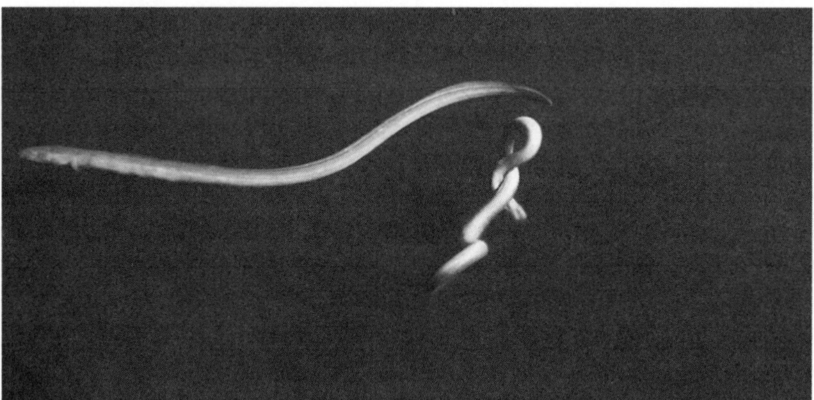

Figure 8. Cutthroat eels, one who is contorted by "toxic shock" in an abyssal brine pool, in "The Deep," from *Blue Planet II.* Still from *Blue Planet II,* "The Deep."

Then, however, we leave the horrors behind to ascend to the ethereal stars by way of something we have heard before: that extremophiles in the vents mean that "there could be life in outer space." The biological abundance and the astonishing beauty and diversity of deep-sea life are left down below as we dream of life far, far above this ultimately creepy, grotesque, threatening, dark place. The episode is, no doubt, deeply disturbing because of its sensationalized othering of abyssal life, articulated with the historical and contemporary discourses of racism.

In Nnedi Okorafor's African futurist novel *Lagoon* (2014), aliens are not deep-sea creatures but beings from another planet who seek to become helpful members of human and more-than-human communities (Figure 9). In this radically inclusive vision, the deep sea is not a separate, horrific, alien zone but instead simply part of the ocean, which is interconnected with the land. The aliens possess shape-shifting powers that align them with the African water spirit Mami Wata.[39] They also possess technologies that can restore the oceans, bring back extinct species, and provide sustainable energy. In contrast with "The Deep," *Lagoon* does not fear darkness, the abyss, or the alien, even as the aliens have transformed some of the marine life into monsters that attack the central characters. Such attacks seem just, given the environmental devastation of the ocean. A decolonial and feminist marine biology is featured in the novel, a science that does not seek separation or domination. Drawing on feminist science studies, Melody Jue argues for what she terms "intimate objectivity," in which "marine biology calmly confronts the elements of the folkloric and the fantastic." "Because *Lagoon*'s scientific intimacy remains open to the surprise of the folkloric and fantastic, aligned with the novum, it constitutes a practice of resistance against western paradigms of scientific practice that are centered around the control and domination of nature based on gendered forms of 'knowing.'"[40] Jue's conception of intimate objectivity, in which marine biology can dodge the domination of nature central to Western, masculinist paradigms, parallels the critiques of the binaries of modern science from Beebe and beyond, as science is capacious enough to embrace philosophical, emotional, and aesthetic matters. Danny, in the novel and the film *Submergence,* can also be read in terms of practicing intimate objectivity, calmly swirling together scientific knowledge with philosophy, aesthetics, and embodied presence.

Returning to *Lagoon,* set in Lagos, Nigeria, we begin with the first chapter, which is written from the perspective of a swordfish: "She slices through the water, imagining herself a deadly beam of black light."[41] The fish seeks to attack "the thing that looks like a giant dead snake" in order to make the ones who "brought the stench of dryness" and the noise, and who "made the world bleed black ooze that left poison rainbows on the water's surface" depart "for good."[42] The "poison rainbows" indicate how petrocolonialism corrupts not only ecological but also aesthetic worlds, such as the "lost paradise" of the coral reef, a "giant world of food, beauty, and activity," which was "blue, pink, yellow, and green, inhabited by sea creatures of every shape and size."[43] Oil spills kill coral reefs, which are also suffering from warming waters and ocean acidification caused by burning fossil fuels. The swordfish sees "something even wilder and more alive than her lost paradise," brought into being by beings from "far, far away," who have somehow cleaned the water and gathered together "creatures from the shallows, creatures from the shore, creatures from the deep."[44] The aliens make the swordfish more powerful, asking what she would like and then making her spear "longer and so sharp at the tip that it sings," and making her eyes "like the blackest stone" so she can see "deep into the ocean and high into the sky."[45] The guests from another realm undertake oceanic remediation while equipping marine life with the

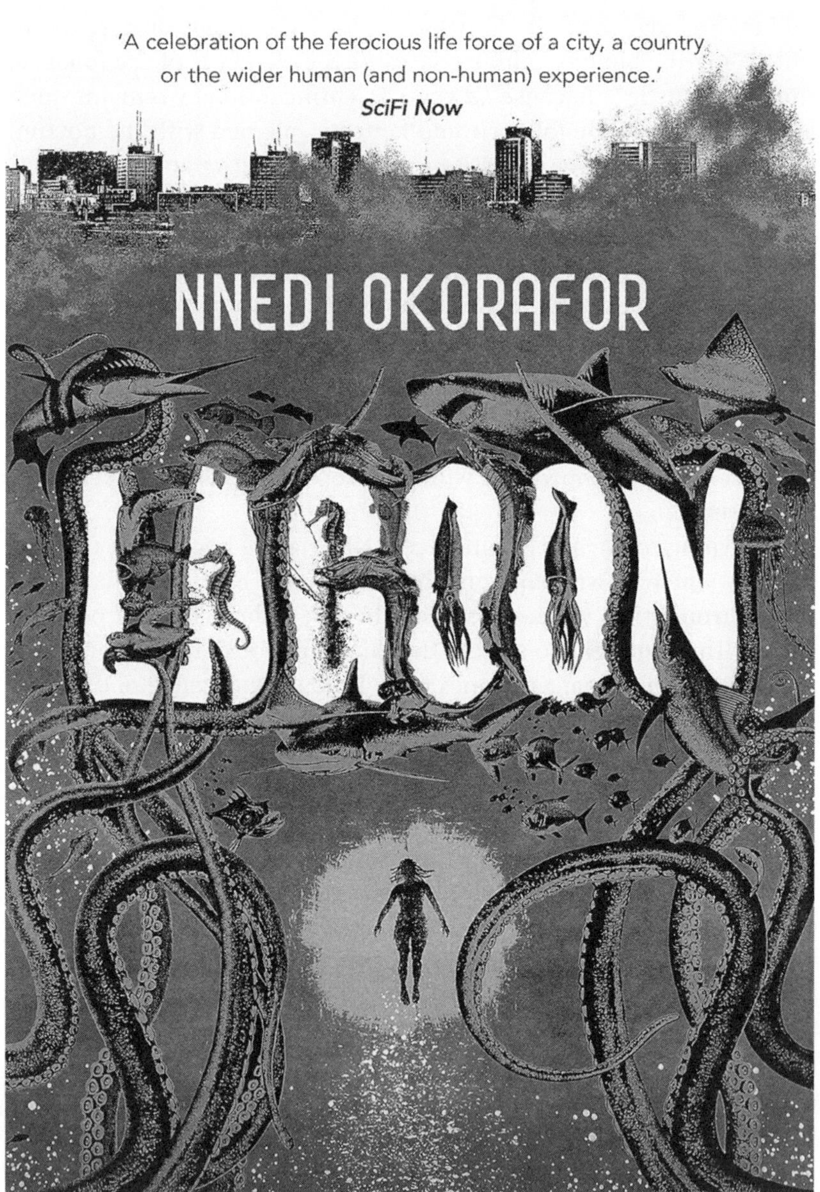

Figure 9. Book cover of Nnedi Okorafor's novel *Lagoon,* depicting revolution brewing in the seas.

capabilities they need to protect themselves. Reminiscent of Octavia Butler's Xenogenesis trilogy,[46] *Lagoon* advocates for generative modes of xenophilia, in which aliens initiate planetary revitalization.

Along with gathering marine animals from various zones, the novel brings together Adaora, a marine biologist; Anthony, a rap star; Ayodele a water witch/alien; and Agu, a soldier, who all play a part in the revolution. The rainbow reappears later as a group of queer people organize under a "great big rainbow-colored sign with a giant black nexus painted in the center."[47] The fabulous queers, wearing headbands with "alien antennae bobbing from them," welcome the extraordinary beings as allies. The aliens themselves reject this term, however, hacking technology to digitally announce, "We are guests who wish to become citizens."[48] While the guests provoke unlikely gatherings and alliances, the novel also suggests that ocean life deserves its own space, as well as sovereignty. This happens dramatically when the central cohort of characters, including the president, are attacked by tentacular monsters. When Agu ponders why "the entire ocean had decided to come after them," he already has the answer: "They were human. They didn't belong here in the deep. . . . Best to leave these waters to the ocean animals and the aliens."[49] Tabea Wilkes argues that the "resolution of the book, therefore, is not that humans have been attacked by aliens, but merely that aliens have empowered the water to remove humans from a space to which they have brought destruction, and to which they perhaps did not belong in the first place."[50] Near the novel's end, the president urges people to stay away from the ocean for now, to embrace the aliens as brothers, and to look forward to a future in which destructive technologies would be supplanted by new technologies, given by the aliens, that would restore biological diversity: "Extinct creatures would return and new ones would appear. Nigeria would have much to give the world."[51] Okorafor's African futurist vision positions Nigeria as the epicenter of global environmental transformation, a revolution that begins with a swordfish's aesthetic appreciation for lost beauty and vivid life in the sea. Granted, this novel is more capacious and nuanced than this reading can begin to suggest. Granted, the novel does not feature a specifically deep-sea aesthetic. Yet it is important to include it here—not only as a corrective to the Eurocentrism and Western "universal" visions that reinforce colonial histories but also to underscore the brilliance of this imagined future in which oceanic monsters and aliens are vital parts of the revolution, existing in their own worlds and inciting more extensive ecological reclamation from the ravages of petrocultures, colonialism, and capitalist extraction (the very monsters that lurk in plain sight). Cajetan Iheka discusses *Lagoon* as an African futurist novel that searches for "alternative spaces for black bodies and other life forms, human and otherwise," recognizing that "the future promised by the capitalist present is antithetical to the promise of black life and warrants an alternative."[52]

The radical promise of Black life is envisioned by Alexis Pauline Gumbs in *M Archive: After the End of the World,* which presents a "black feminist metaphysics" that boldly exceeds all modes of disconnection: "there is no separation from the black simultaneity of the universe also known as everything also known as the black feminist pragmatic intergenerational sphere, everything is everything."[53] Gumbs envisions a profound and expansive Black feminist universe that supersedes individuals as well as the human as such, saturating everything with "the reality of the radical black porousness of love."[54] This cosmic vision contrasts with that of the racist "they" who "had this thing about darkness, the bottom of the ocean, outer space. they were afraid of it, they wanted to penetrate it."[55] Such twisted impulses of fear, desire, and domination may recall the toxic shock scene in *Blue Planet II.* Gumbs weaves together feminist critiques of Western scientific objectivity (and its metaphorical penetrations of nature) with denunciations of the terror of the black void. She departs from this sort of science, recording instead, in the "lab notebooks" section of the volume, descriptions of the "black oceanists" who "trained themselves and each other not to be afraid of going black (that was what they called it) for days at a time."[56] "Going black," for the oceanists, entails kinship, evolution, and acceptance of the abyssal realm. "They were not afraid to slow and evolve their breathing. they were not afraid of their kinship with bottom crawlers who could or could not glow. they were not afraid of being touched by what they could never see, never bring back to the light, never have a witness for."[57] There are no aliens here—nothing strange, nothing creepy, nothing monstrous or abject. Intimate modes of knowing alter the knowers, who evolve through their kinship. They do not dominate through distancing but are instead being touched by what they could "never bring to the light." The ability to allow oneself to be touched without having to name, know, circumscribe, or repel what is touching you is a strong, brave, and porous mode of being, which flows with and becomes with other lifeforms. We could read this as an aesthetic of the haptic, being touched as a beautiful transformation into as yet unknown modes of being that are submerged within vibrant darkness as "the reality of the radical black porousness of love." Gumbs's poetic, visionary prose is itself an aesthetic marvel, crafted through intermingling Black feminism with scientific information, including how calcium and magnesium catalyze bioluminescence. Abyssal beauty not only glitters but profoundly, elementally, interconnects, such as when the "critical black marine biologists, scientists of the dark matter under fathoms, suggest that there may be a causal relationship between the bioluminescence in the ocean and the bones of the millions of transatlantic dead."[58] This is not an aesthetic that can be commodified, consumed, and dismissed but instead is something deeply and powerfully meaningful and transformative. Gumbs's revolutionary sense of love reveals, even at a literally elemental level (calcium and magnesium), that the horrors of human history and the life at the bottom of the sea cannot be severed and should not be forgotten. Less violent, less dominating, and less distancing modes of knowing and being are possible.

Gumbs's experimental, poetic, speculative storytelling incorporates history, science, theory, and Black feminism in order to transform present realities by way of potent modes of kinship and radical interrelation. In *Dub: Finding Ceremony,* Gumbs advocates for a capacious, welcoming, and generative model of kinship. She explains that when she began to listen to her ancestors, she included "speakers who have never been considered human." She explains: "And while that category of the never-considered-human tragically includes my enslaved ancestors, my disabled ancestors, my queer and indigenous ancestors and everyone subject to the police radio codes Wynter writes about in *No Humans Involved,* it also generatively includes whales, corals, barnacles, bacteria and more. Kindred beyond taxonomy."[59] The *Blue Peril* video with which this chapter began, which voices opposition to deep-sea mining from an Oceanic enunciation, also articulates kindred beyond taxonomy. Kinship, which interrelates one's very being with the being of others, is antithetical to the quotidian, unremarkable ontologies, orientations, and systems that routinely render living beings into resources for use or disregard them as mere collateral damage. Using the term "kindred," not "kinship," Gumbs moves away from kinship systems, a staple of anthropology, and toward beings connected by genealogies, alluding, perhaps, to Octavia Butler's novel *Kindred,* in which time travel enables present-time reckonings with transatlantic slavery. "Kindred beyond taxonomy" suggests inclusive multispecies relations of respect and responsibility, speculatively rich and tangible relations that refuse systems of categorization that sever humans from other beings.

The Abyss—Too Close at Hand?

Throughout this study, I have invoked the abyss at hand to signify the strange way in which creatures from the depths circulate though modes of intimate mediation that spark aesthetic encounters. Such encounters may unmoor the distancing, rational frameworks of scientific objectivity; inspire a sense of curiosity and awe for extraordinarily heterogenous abyssal life; and provoke speculative metonymic mappings of the lifeworlds of these creatures. Such global visions, which expand what an ostensibly universal environmental subject must care about, echo imperial and colonial mappings, yet also seek to disrupt capitalist and neocolonialist destruction of marine ecologies. Because both taxonomy and environmentalism are implicated in racism and colonialism, for those of us implicated in settler colonial histories, there is no innocent ground from which to launch deep-sea conservation; yet anyone who lives within capitalism and petroculture is always already complicit with the destruction of marine ecologies, which proceeds offstage, largely ignored. But as highly aestheticized captures of abyssal creatures—swimming up to greet us in Else Bostelmann's surreal paintings, framed as treasures on digital Census galleries, and vibrantly staged as extraordinary beings hovering within Nouvian's book—provoke a sense of immediacy or presence, the constitution of the world shifts. Images transport the shape, form, colors, and distinctiveness of these astonishingly heterogenous abyssal animals to the screen in front of us or to the page, positioning them as seemingly at hand. Immediacy, as it refers not only to space but also to time, resonates with the urgency of deep-sea protection.

As deep-sea animals circulate on computer screens, books, social media, art, video games, and film, their seeming immediacy may be part of larger cultural trends. Anna Kornbluh's *Immediacy, or The Style of Too Late Capitalism* presents a provocative diagnosis of this moment, when "ecocide has already taken place." She grants that "the ideology of immediacy holds a kernel of truth; we are fastened to appalling circumstances from which we cannot take distance, neither contemplative nor agential, every single thing a catastrophe riveting our attention."[60] She contends, however, that "immediatism is a reaction to a crisis that fails the bar of strategy, a reflex that is ultimately crisis-continuous."[61] Kornbluh defines "theory" in ways that sever us from corporeality, materiality, affect, intimacy, kinship, interrelation, and ecological enmeshment, reinstating transcendence as the route to truth: "Theory effects distance, abstraction, movement away. Immediacy foments intimacy, immersion, the negation of intercession. Theory takes us out of a situation, out of phenomenality, out of ourselves, and into realms of reflection that escalate to include the dislocation of our ineluctable situatedness, and speculations about inexperienced possibilities."[62] The movement that "theory" would effect, toward "distance" and "abstraction," is complicit, however, with rendering lively ecological worlds into inert resources for capitalist enterprises, such as Craig Venter's bioprospecting, which transforms life into code, discussed in chapter 3. Rather than reinstating vertical structures that endorse abstraction and transcendence as the route to truth and objectivity or advocating for theory as concrete or confined to immediacy, I have argued throughout this book and elsewhere for modes of thinking that oscillate, suspended between these poles—for example, in material epistemologies that I call "thinking as the stuff of the world"; in transcorporeal contemplations that situate the self as substantially embedded within the material agencies of this postnatural age; and for the practice of scale shifting from embodied vantage points within the Anthropocene rather than viewing the wreckage from some imaginary position in the sky. Here in this book, considering the abyss at hand means to encounter abyssal life in ways that seem intimate but depend on scientific, technological, and artistic mediations. The aesthetic in such encounters can act as a lure to attachment and speculation about creaturely lives in the abyss. Unlike many companion species, abyssal life is never truly at hand, yet scientific and aesthetic captures and currents of mediation may provoke a sense of connection that encourages publics to extend environmental concern to the seafloor. "Theory" has been a contested term in academic culture, at least since the 1990s. With the arrival of cultural studies, as well as the mushrooming of fields such as environmental humanities, feminist and queer theory, Black studies, Indigenous studies,[63] and multispecies studies, limiting theory to a mode of abstraction circumscribes and excludes, casting out the bold and intrepid work of theorists who see the binaries of Western thought—mind/body, reason/emotion, human/animal, culture/nature, material/abstract—to be the very thing that is crisis continuous. Shaped by these hierarchies, the epistemologies of distancing and abstraction have been foundational for capitalism, colonialism, racism, sexism, and the domination of all of those who have never been considered human.

Many of the aesthetic encounters with deep-sea life, especially since the Census of Marine Life, could certainly be critiqued as an immediacy that "crushes mediation."[64] I see such moments as shallow, consumerist reactions that take what they enjoy from the images, participating in a long history of objectifying and consuming nonhuman animals as flesh or entertainment. Actual immediacy is, of course, not possible for most encounters with abyssal life. As Melody Jue aptly puts it, "Rather than being immediately present to us, the (deep) ocean emerges as an object of knowledge only through chains of mediation and remote sensing measurements that allow us to build up imaginative pictures of what life in the ocean is like, often from points of view that would be physically impossible to inhabit without the protection of a submersible."[65] Such imaginative pictures involve scientific and aesthetic captures, made possible by labor, funding, and technologies, but inspired by curiosity about species and concern for ecologies and lifeworlds. This is not a matter of the sort of representation that Kornbluh describes, which divides nature from culture and reality from representation, but instead involves chains of mediation in which the image of a deep-sea animal presents a lure for rumination, with musings that intimate something of this particular animal, a "circulating reference"[66] that retains some aspect of the creature itself. Such musings are mediated by cultural tropes and aesthetic categories, yet contemplating even just the basic morphology of an animal as it arrives on a screen can evoke a potent biophilia, a moment of encounter with the undeniably marvelous singularity of this species—a species that has somehow happened, has somehow come to be, within long, roiling evolutionary histories.

Aesthetic encounters inspire pleasure, awe, and joy as they insist on the inherent value of the diversity of animal being. Even if—especially if—speculations about other species never arrive at a final destination or settle on some sort of truth, the practice of imaginatively hovering with abyssal life encourages more capacious modes of concern. Once you have seen a dumbo octopus or a sea angel—Google them now—how could you cast the deep-sea biome outside the parameters of what deserves protection from rapacious plunder?

The posthumanist sense of awe and radical unknowing that the multitude of heretofore unseen, stunning, and extraordinary abyssal species can provoke may, however, collapse into itself when the animals are too close at hand, too immediate. Although I was excited to see Monterey Bay Aquarium's 2022 exhibit, *Into the Deep: Exploring Our Undiscovered Ocean,* after many years of working on this book, the actual experience was jarring. After my silent, solitary musings on the gorgeous photos in Claire Nouvian's collection, discussed in chapter 3, I was taken aback by the crowded, noisy, raucous exhibit. When I told the staff that I was writing a book on the deep sea and had requested to meet with one of the exhibit designers to ask them a few questions, they informed me that I would not be allowed to ask anyone any questions. The staff, with their walkie-talkies, tracked me down, watching me with suspicion while I obediently moved through the entrance line. What was going on? Inside the exhibit, people more important than a professor—big donors, perhaps?—were treated to special tours. In an era when billionaires can descend to the depths just for fun, without furthering science or conservation, privileged access to deep-sea life on display is not surprising. The exhibit itself featured large video displays, models of deep-sea species in dioramas, and columnar tanks of live animals, including many tiny jellies—lobed comb jelly, bloody belly jelly, purple lipped jellies, abyssal combed jelly, as well as a common siphonophore. Another large tank featured predatory tunicates, droopy sea pens, deep-sea carnation corals, armored sea cucumbers, glowing sea cucumbers, and mushroom soft corals. Some of the tiny creatures were enchanting in inverse proportion to their size, such as the newly discovered diminutive red jellies, the size of a wild blueberry, glowing in their tank.

Watching the jellies pulse, being so close to abyssal life, was enchanting, even though I have long been critical of putting living animals on display for entertainment and I do not frequent zoos or aquariums. I made an exception for this book, but I cannot deny my curiosity and initial excitement. Encountering abyssal life pulsing within small tanks was poignant, mesmerizing, breathtaking. Yet I was reminded of the futility of the encounter, encapsulated in John Berger's famous quote: "The zoo to which people go to meet animals, to observe them, to see them, is, in fact, a monument of the impossibility of such encounters. . . . You are looking at something that has been rendered absolutely marginal; and all the concentration you can muster will never be enough to centralise it."[67] Metonymic with their habitat, lifeworld, and biome, deep-sea life displayed in arid, terrestrial structures cannot help but disappoint, despite the hefty admission fees. The animals, removed from their vast lifeworlds, become merely animals on display. Eva Hayward, in her essay about visiting the jellyfish tanks at Monterey Bay Aquarium, asks, "If captivity is always mediated . . . how can we see mediation as a dynamic of encounter, even an ethical one?"[68] Hayward provocatively suggests that we "encounter the conditions of encountering 'real animals,'" not "necessarily contact or direct meaning but rather a sensuous rapport or energetic cadence."[69] Hayward's phrase, "encountering the conditions of encountering," suggests the sense of encounter I have been attempting to describe—an encounter that is virtual but nonetheless affecting. Her gorgeous phrasing resonates with an embodied, sensual aesthetic—a perception of and convergence with the sight of the animal, and an energetic alignment with the pulsing rhythms. Raptly peering into the columns of pulsing jellies, I felt such a sensuous rapport. I was in awe of their diminutive, wispy beauty and their pulsing rhythms. However, their containment and display severs the animals from their abyssal lifeworlds, enlisting them as novelties, entertainment, or living educational exhibits. It was sad, frankly, to see living beings that hover in distant worlds—unfathomable places that inspire musings on the limits of human understanding—contained within small cylinders, seemingly competing for shreds of people's distracted attention. The abyss may be most captivatingly at hand, paradoxically, when visual captures circulate as highly mediated aesthetic images, leaving the living creatures in the depths rather than making them captives on display.

Exiting the exhibition, viewers stroll by windows looking out onto Monterey Bay, allowing them to imagine the species on display inhabiting the actual ocean. In this way, the exhibit supports the practices this book has discussed: the oscillation between an intimate aesthetic encounter and a speculative mapping of creaturely life and habitats, especially as they are saturated with anthropogenic harms. Some elements of the exhibit nod toward ecological concerns. The short making-of film underscores the exhibit's achievement: displaying animals that have never before been put on display in order to foster "awareness."[70] The exhibit also included a video game in which players must try to determine, as the animals would, whether something that seems like food is indeed edible or is actually a plastic bottle, bag, or straw, offering a primer on the long reach of human impacts. The inclusion of the game is helpful, if perfunctory. The game adds to the provocative weirdness of the exhibit itself. As deep-sea life pulses in tanks, far from home, chemical and plastic pollutants invade even the animals in the deepest trenches while plastic bags leave trails across the seafloor.[71] The living specimens are contained; the pollution is not.

The provocative implosion of human venues and benthic life reaches a crescendo with the touch tank. Including a touch tank in the exhibit intensifies the immediacy of the encounter by shifting from the visual to the haptic, ostensibly removing barriers to connection. As children lined up to pet the giant deep-sea isopod in the touch tank, I wondered what sort of cruel love this was, subjecting these living beings to the strangeness of human touch within the waterless, bright, bustling, noisy, terrestrial world. How could they not be suffering? When they weren't being handled, did they have room to move about? Benthic isopods are not companion species;[72] they did not coevolve with humans. It is unlikely they would consent to being captured, brought to the surface, and handled by children. The haptic immediacy of the touch tank could epitomize the problematic Kornbluh has diagnosed: "immersive intensity transmits an instant message; it brooks no abstract mediations."[73] Indeed the touch tank aims to transmit instant experiences if not immediate messages. It's hard to imagine what sort of message an isopod would send in this jarring, insupportable situation. Nonetheless, the touch tank scenario itself and the now-iconic weirdness of deep-sea life mediate the encounter for aquarium visitors in ways that are visual, spatial, embodied, and institutional rather than abstract. The physical encounter with an animal from the seafloor, which could disturb human presumptions about the contours and geographies of planetary life, is already naturalized within the family-friendly genre of the touch tank, relieving children and adults of the effort it would entail to ponder how the heck a giant deep-sea isopod came to be on display, and what that animal could possibly be experiencing. The making-of film about the exhibit promotes the idea that the deep sea is virtually unknown, while the touch tank compresses and domesticates the ostensibly unknown, offering an impossible encounter with an animal conveniently rendered as a prop or object, radically dislocated from its world.

By contrast, the alluring close-up photo of a giant isopod of the deep sea on the Ocean Conservancy's website shows the animal digging its claws into the sediment, gorgeously going about its life. The story, "Meet the Giant Isopod of the Deep Sea: Get to Know and Love the Giant Roly-Poly Bugs of the Ocean Floor," promises knowledge paired with loving encounters.[74] The inserted video from National Oceanic and Atmospheric Administration begins by showing the isopod swimming into view just above the seafloor, a small critter moving across a wide benthic terrain. The camera gives the isopod plenty of space, on and above the seafloor, an utterly inhuman habitat, before zooming in. The voice-over notes the charisma of these creatures and their ecological importance as benthic scavengers, concluding, "To some people they are really cute; to some they are the stuff of nightmares." Dismissing human judgment as inconsequential, the video endorses a disanthropocentric, scientific attention to the animal itself. Jennifer Fay, drawing on Sigmund Kracauer's cinema theory, argues that even as images are created for human spectators, "the human is potentially just another thing among things, subject to the technology's indifference."[75] She asks, "If, in the image, humanity enjoys no necessary priority over other animate and inanimate matter, by what right do we humans claim possession of the world and submit the planet to our designs?"[76] By what right indeed? Forcing the planet to "submit to our designs" characterizes the Anthropocene, which, as Arturo Escobar argues, results in the ravaging of beauty itself.[77] Because intimacy with another being respects the integrity and difference of that being, intimate mediations do not require contact, but instead create paths of connection with creatures who can be themselves, within their own worlds.

◈ Epilogue ◈

Who Cares?

Calculating the Value of Abyssal Life

> We are suspended over the abyss of the impending absence of nature, but this suspense, in turn, gives way to exhilaration and movement.
>
> —Bonnie Mann, *Women's Liberation and the Sublime*

In 2020, marine biologist Alan Jamieson, along with Glenn Singleman, Thomas D. Linley, and Susan Casey, published "Fear and Loathing of the Deep Ocean: Why Don't People Care about the Deep Sea" in the *ICES Journal of Marine Science,* which begins with the Charles Saxon cartoon with which this book begins. They describe a large scientific conference with two hundred delegates from the United States and Europe that was "supposed to focus on the practicalities of launching a round-the-world deep-sea exhibition." They note, however, that an "unexpected trend appeared in presentations": the repeated question, "How do we get people to care about the deep sea?"[1] Interestingly, the authors pose this question of caring as a scientific failure rather than a cultural or media failure, asking how the work of the "deep-sea scientific community" is "going wrong, if, in 2020 we are still asking ourselves why don't people care about the largest ecosystem on the planet?"[2] To answer this question, they discuss thalassophobia, deep-sea monsters across history and cultures, the vertical dichotomies that demonize the depths and deify the sky, the fact that the public only encounters deep-sea life through digital media, the misleading idea that science knows little about the depths, as well as aesthetics, ethics, and values. While I agree with many of their arguments, after having pondered similar questions for over a decade I am taken aback by the contention that in a "deep-sea context we are dealing with species that are not portrayed with a particularly aesthetic value."[3] Really? I'm not sure what to make of that statement, given the many examples throughout this book that illustrate how abyssal life has been displayed in highly aesthetic modes, from the surreal to the lavishly gorgeous to the vibrantly stunning. Claire W. Armstrong et al. refuted the essay in "People Do Care about the Deep Sea. A Comment on Jamieson et al.," arguing that Jamieson and his coauthors merely assume people do not care, assembling evidence from art, literature, film, and popular science writing, along with social science surveys of what various peoples value, citing a global survey that showed 81 percent of people do care about the deep sea, even though that care has not resulted in environmental protection.[4] Jamieson, replying to the reply, this time with Thomas D. Linley and Prema Arasu, analyzes the evidence included in the response of Armstrong et al., arguing that some of the sources in Armstrong's reply are not actually about the deep sea and others promote the fear of the deep. Jamieson and his colleagues conclude by stating that the original "Fear and Loathing" essay "challenged the way in which the public perception of the deep sea is based on sensationalist misrepresentation and outdated quantifiers of how much we know, or don't know, about the largest habitat on earth." Noting that the ocean inspires "fascination and wonder," they argue that "these emotions tend to take a negative turn as depth increases and light decreases."[5] Most notably for our purposes here, the aesthetic is ignored in this final response. It vanishes as a potential force for eliciting care, connection, and concern. While a deep-sea creaturely aesthetic has, to me, seemed a surprisingly persistent and potent spark for luring publics into speculating about abyssal worlds, it can be readily dismissed as diffuse, undefinable, nebulous, and thus inconsequential. How can the aesthetic be located, defined, measured? The aesthetic eludes definitional capture and quantification. On the third episode of the "Deep Sea Podcast," by Jamieson and Linley, entitled "Aesthetics of the Deep Sea with artist Alex Gould," however, Linley asks, "Are we getting this wrong? Are we coming at this from too much of an analytical scientific angle? We need to look at the aesthetics of these animals and how people engage with them." I hope that this book, which has attended to the aesthetics of deep-sea life across science, scientific memoir, science fiction, art, film, and popular media, has contributed an array of analyses, interpretations, and speculative arguments for what a creaturely aesthetics can do.

The creaturely aesthetic is nearly absent within Ned Beauman's near-future extinction satire, *Venomous Lumpsucker* (2022). As the velocity of anthropogenic extinction escalates in Beauman's dystopia, most animals are already gone, nearly gone, or "existing" solely in digital or ghastly modes: "No ringtails anymore, no harbor porpoises, no velvet scoters, no European eels, no angel sharks, and practically no venomous lumpsuckers."[6] Lumpsuckers are a family of small cold-water benthic fish, some of which can live at depths of five thousand feet. Despite their unappealing common name, the Ocean Conservancy, in "Meet the Lumpsucker," quips that "you can't deny they're pretty stinking cute!"[7] Yet the aesthetic appeal of the lumpsucker, and other near-extinct species, is not accounted for in this complicated novel. The central conflict counterposes Mark Halyard, a mining executive, and Karin Resaint, an animal cognition scientist, within a global biopolitical regime in which industries or states engaged in mining or other environmentally destructive exploits must pay "extinction credits." Beauman dubs this the "extinction industry," a gory phrase telegraphing how industrial capitalism essentially produces the death of species. The novel's extreme version of biodiversity banking means that if countries, industries, or individuals destroy species, they must pay, at fluctuating market rates. Not surprisingly, the market has more mobility and dynamism than the scattered remnants of nonhuman life. At the same time, other entrepreneurs profit from cramming together species that would not inhabit the same ecosystems, or worse, storing specimens or uploading brain scans in biobanks, with delusional hopes that someday those species can be reanimated. However, as Halyard points out, "nobody cares about most of these species. . . . So once the poor little fucker gets scanned, that's it. That's the end. It's just going to sit there in the database forever like some moldy old library book that nobody will ever read."[8] Why bother to read scans of species that have no habitats or ecosystems to inhabit? As the oceans rise, the "vanishing Pacific kingdom of Tonga" placed "their entire ecological heritage in the fumbling hands" of the scamming biobanks, with tragic results.[9]

In this global capitalist biopolitical dystopia, it is animal intelligence that makes some species more valuable than others—precisely thirteen times more valuable, in terms of the cost of extinction credits.

The number thirteen is not only unlucky but also weirdly exact, amplifying the absurdity of exclusively valuing the ability to calculate and reason in abstract modes as intelligence. When Resaint, hired by the marine mining company to certify that the lumpsucker is not intelligent, discovers that it is "one of the most intelligent creatures on the planet," drama ensues.[10] Her employer pressures her to ignore her findings because so much money is at stake. The lumpsucker's intelligence is described in capitalist terms. As a cleaner fish, a single lumpsucker "could attend to over a thousand clients" in a day, requiring, in this risky relationship of trust, that the fish keep a "detailed mental database of all its clients."[11] The discourse of clients and databases is all too human, capturing this creature within the homogenizing, flat Anthropocene. Given such talents, it makes sense that when tested, the lumpsuckers "recognize[d] themselves in a mirror" and performed "better than chimps on certain logic puzzles."[12] Imagining a fish who dwells in dark aquatic zones being subjected to such intelligence trials conjures up the entire history of such experiments, leaving readers to ponder the discomforting scenarios in which particular species have been forced to step up and prove their worthiness. It is fitting that the economic system in this dystopia values intelligence, however, given that abstract, rational thinking is consonant with the transformation of use value into exchange value, as well as with global financial systems, which proceed through innumerable calculations that care not for humans or more-than-human life. While Resaint recertifies the lumpsuckers in accordance with the system, and advances her own plot that requires their intelligence (no spoilers here), she believes, more deeply, in the inherent value of every species that has evolved: "What did it matter if anyone appreciated it, if it did any good for anyone?"[13] Significantly, however, when Resaint tries to convince Halyard that other species possess inherent value, she creates a speculative thought experiment that is saturated with a rich, lively, creaturely aesthetic.

> Resaint asked Halyard to imagine a planet in some remote galaxy—a lush, seething, glittering planet covered with stratospheric waterfalls, great land-sponges bouncing through valleys, corals budding in perfect niveous hexagons, humming lichens glued to pink crystals, prismatic jellyfish breaching from the rivers, titanic lilies relying on tornadoes to spread their pollen—a planet

> full of complex, interconnected life but devoid of consciousness. "Are you telling me that, if an asteroid smashed into this planet and reduced every inch of its surface to dust nothing would be lost? Because nobody in particular would miss it?"[14]

Aesthetically potent, featuring (land) sponges, corals, and prismatic jellyfish, the fictional planet glitters with marine beauties and lush, enmeshed life that is "devoid of consciousness." This refutation of the idea that species should be valued in terms of their intelligence—and, more broadly, in terms of their use for humans—turns on a speculative aesthetic that is absent from the rest of the darkly comic satire. The lack of attention to the aesthetic within the novel is appropriate, however, as the value of this inestimable, sensory, and embodied mode of recognition, appreciation, and interconnection cannot be calculated in biopolitical or capitalist terms. Neuroscience could measure the effects of beauty, awe, and wonder on human health and well-being, but even that sort of data is qualitatively different from, say, a cost–benefit analysis of deep-sea mining for cobalt that would endanger benthic species. Moreover, the ineffability of the aesthetic allows it to be expansive enough to embrace the immense, kaleidoscopic biodiversity of abyssal species. Mediated encounters with fish touted as ugly, for example, can still evoke curiosity, care, and concern. In other words, the aesthetic opens out into modes of recognition that exceed the parameters of beauty as such. As the aesthetic is ineffable, it is also incalculable.

Philosophy, science writing, critical animal studies, and the posthumanities have debated the question of what makes particular animals valuable—or, in other words, what makes them deserving of human concern. Such assessments have turned on whether nonhuman animals can think, suffer, feel, communicate, recognize themselves, imagine the perspectives of others, care for their young, create music, perform mating dances, create their own cultures, create and use tools, serve as companion species to humans, provide food, fiber, or entertainment for humans, serve as keystone species or ecosystem engineers, or perform other vital ecosystem services. The calculation of the value of various species confines the recipients of such assessments within circumscribing, distancing, and sometimes denigrating frameworks. The aesthetic—wispy and inconsequential as it may seem—suffuses viewers with awe, wonder, desire, empathy, exhilaration, and other cognitive, emotional, and/or embodied responses, which lure us away from human exceptionalism and the capitalist objectification of other beings.

Rather than homogenizing abyssal life or deep-sea creatures into one category, it is vital to speculate across a multitude of singular species, their beings, and their lifeworlds. The multitude of deep-sea species—many of which have only been recently discovered, many of which confound our sense of what an animal can be or do, and many of which provoke an aesthetics of the surreal—invite fabulously divergent speculations. Chemosynthetic life; diurnal vertical migrators; extremophiles; piezophiles; slow, sessile animals of the seafloor; "living fossils" that endure across geological epochs; pelican eels with gigantic mouths; anglerfish sporting their own lures; a multitude of glowing, pulsing jellies; and salp, with their complex life histories that include "sexual reproduction, asexual budding, solitary individuals, and chains of clones"[15]—all these can lure us away from monotonous, flat, manufactured anthropogenic and anthropocentric landscapes of the Anthropocene. Moreover, delighting in the breathtaking biodiversity of abyssal life veers away from xenophobia and toward an embrace of alterity and queer ecologies.

Aesthetic encounters with deep-sea life can evoke mediated modes of intimacy with a kaleidoscopic abundance of life, underscoring that the largest ecosystem on the planet warrants protection. Moreover, the sheer number of vertiginously diverse abyssal species—stunning in their distinctive shapes, forms, colors, textures, and patterns—implicitly declares that biodiversity is vital for environmental visions. Aesthetic encounters with a creaturely abyss, seemingly at hand, may, one can only hope, help avert the destruction of species and ecosystems, a future already too close at hand. Bonnie Mann, describing an encounter with a wild turkey, not an octopus, coral, or siphonore, writes that the "experience of nature is plagued by the terror of nonexistence, the fear that natural beauty will be forever stamped out. . . . We are suspended over the abyss of the impending absence of nature, but this suspense, in turn, gives way to exhilaration and movement. We are exhilarated. . . . Such beauty still exists!"[16] Aesthetic recognition may seem too delicate, too frivolous, or too ineffectual to provoke the immense ethical, political, and economic transformations necessary in this moment of accelerating extinctions, but how else can the jellies, squid, siphonophores, hatchetfish, lanternfish, isopods, barreleye fish, harp sponges, anemones, feather stars, fire stars, corals, octopuses, tubeworms, dragonfish, sea pigs, and lumpsuckers catch the attention of those who need persuading? Shimmering, glowing, gliding, darting, undulating, pulsing, and fluttering—with or without eyes—the abyss stares back, in suspense.

Acknowledgments

This book, long in the making, has been a labor of love, enabling me to read widely across something some of us now call the blue humanities and to become captivated by a multitude of extraordinary and gorgeous creatures. The most thrilling part of the research was undoubtedly the time I spent in the archives. I am extremely grateful for a library research grant from the Friends of the Princeton University Library, which enabled me to study the William Beebe Papers in Princeton's Special Collections. I am also grateful for the summer research grant I received from the University of Texas at Arlington to study the art of Else Bostelmann in the Wildlife Conservation Society's archive at the Bronx Zoo. Madeleine Thompson, director of the library and archives at the Wildlife Conservation Society, was a generous and extremely knowledgeable guide. I am also grateful for Brett Dion's assistance with securing Bostelmann's art for publication. I was honored to receive the Presidential Fellowship in Humanistic Studies at the University of Oregon for this project, which supported the illustrations, permissions, and a research trip to a deep-sea exhibit. Thanks as well to the Oregon Humanities Center for a subvention grant that helped with the cost of the illustrations. I appreciate Christina Lujin's skillful work processing the funding. I would also like to thank Franny Gaede, director of digital scholarship at the University of Oregon libraries, and Peter Hanlon, director of public engagement at the University of Rhode Island graduate school of oceanography, for attempting to bring some of the images from the Census of Marine Life website back to (digital) life. I am grateful for the expert assistance of Zenyse Miller at the University of Minnesota Press, who guided me through the permissions, illustrations, and other processes, and to the entire team of talented people at University of Minnesota Press who ushered this book into the world. Finally, colossal thanks to Doug Armato at the University of Minnesota Press, and Cary Wolfe, editor of the Posthumanities series, for their enthusiastic support of this project and their patience with its long development. One of my first presentations on abyssal aesthetics was during an ASLE (Association for the Study of Literature and Environment) plenary with Cary Wolfe, so it seems especially propitious that this book would appear in the Posthumanities series.

I have been quite fortunate to be part of vibrant academic communities in which ideas and fields emerge and develop. My thinking about deep-sea life has been provoked and enriched by invitations to write and speak. I wish I had the space and the memory to thank everyone who discussed this project with me. There are so many scholars for whom I feel such admiration and affection; it is not possible to list them all here. Some of the most generative and fabulous academic events I've ever attended have been those arranged by Astrida Neimanis and Cecilia Åsberg, both together and separately; I remain inspired by their visions of the feminist posthumanities. I am also grateful to Rosi Braidotti for her engagement with my work and her generous articulation of feminist posthumanism. Immense thanks to Jeffrey Jerome Cohen for several invitations to write and speak about the blue humanities, and for supporting my work more generally. I've given too many talks on deep-sea animals to include here, but I'd like to express my gratitude for a few of the most rewarding events: serving as lead scholar for the Colby Summer Institute, being the guest scholar for the International Association for the Fantastic in the Arts (thank you Sherryl Vint), participating in the Anthropocene Feminism conference (thank you Richard Grusin), and presenting at Babel on the Beach, where we bodysurfed in the morning and swam in the afternoon with Vanessa Daws, a swim artist! I've also presented my deep-sea scholarship on podcasts and virtual talks for Green Dreamer, OceanUni, and the Institute for Postnatural Studies. Thank you! Finally, even this brief account of appreciation would be remiss if it did not extend to the inspirational work of scholars (and an artist) in the blue humanities: Irus Braverman, Margaret Cohen, Liz DeLoughrey, Megan Hayes, Elizabeth R. Johnson, Melody Jue, Stefan Helmreich, Lydia Lapporte, Elizabeth Burmann Littin, Steve Mentz, Astrida Neimanis, Sue Reid, Nicole Starosielski, and Marina Zurkow. Ursula Heise and Cate Sandilands have been friends since the Boston ASLE; I am grateful for their brilliant work and their friendship. I'm also grateful for all my longtime as well as more recent interlocutors and friends in the ASLE and SLSA (Society for the Study of Literature and the Arts) communities.

The writing of this book has extended across two different lives, one in Dallas, Texas, and one in Eugene, Oregon. I miss my fabulous colleagues and friends from Dallas, including Rajani Sudan, Chris Morris, Penny Ingram, Cedrick May, Susan Hekman, Wendy Faris, Neill Matheson, Justin Lerberg, Jason Hogue, Amy Tigner, Ken Roemer, Laura Kopchik, and Tim Richardson. I continue to be inspired by the work and activism of Priscilla Solis Ybarra. I've learned so much from the dynamic duo of Christy Tidwell and Bridgitte Barclay, and I value their friendship. I deeply appreciate all the animal people, including Cary Wolfe, Lucinda Cole, and Ron Broglio, and am inspired by Ron's experimental approach to the humanities. Kai and Felix made the cross-country trip with me; Heidi became part of the family during the pandemic. My aunt Janet moved to Oregon and became my hiking pal. I am extremely grateful for my Oregon friends, especially, Faith Barter, Louise Bishop, Liz Bohls, Lara Bovilsky, Ashley Cordes, Caleb Connolly, Katie Lynch, Tres Pyle, Emily Scott, Kelly Sutherland, Sarah Wald, Betsy Wheeler, and Mary Wood. Cheers to the entire exuberant environmental studies community, with our very happy happy hours, and especially to Mark Carey, Lauren Hallett, Nicolae Morar, Erin Moore, Lucas Silva, and Richard York. I've been fortunate to work with brilliant graduate students at the University of Oregon, including Katrina Magiulli, Nate Otjen, Lisa Fink, Kathleen Gekiere, Megan Hayes, Moe Gámez, Dakota Brown, Shane Hoyle, and Genevieve Pfeiffer. Immense affection and gratitude to my dear friend Michael Allan—our river walks are everything!

Finally, I'd like to extend heaps of love and gratitude to my mother, Charlotte Fisk; my sisters, Kathy Hoglund and Carolyn Hohman; and my dear friends Karen Sue Coates, Mary Hocks, Susan Hekman, Rajani Sudan, and Chris Morris. Cascades of love flow—always—to River Alaimo and Kai Engwall! This book is dedicated to Kai, who, appropriately, not only possesses a strong aesthetic sense but also eagerly plunged into snorkeling at four years old. May we visit more museums together and spend more time in the ocean!

Notes

Introduction

1. Saxon, untitled cartoon; Koslow, *Silent Deep,* front matter.
2. Van Dover, *Deep-Ocean Journeys,* 160.
3. Mukerji, *Fragile Power,* 150.
4. Scales, *Brilliant Abyss,* xii.
5. Widder, *Below the Edge,* xvii.
6. Widder paraphrasing Mark Watney, 302.
7. Widder, xx.
8. Jamieson, Linley, and Arasu, "Reply To," 2.
9. Deep Sea Conservation Coalition website, https://deep-sea-conservation.org/explore/.
10. Starosielski, "Depth Mediators," 262–63.
11. Mawani, *Across Oceans,* 57.
12. *Meg 2: The Trench.*
13. I use the term "conservation" throughout this book in invisible scare quotes, given its problematic histories, including its links to classism, racism, colonialism, white supremacy, and man making. Systemic problems cannot be resolved with terminology alone; moreover, I do not wish to whitewash the deleterious histories of environmentalism and conservation movements, many of which I have written about previously.
14. Rozwadowski, *Fathoming the Ocean,* 137.
15. Rozwadowski, 15.
16. Steinberg, *Social Construction of the Ocean.*
17. Warren, *Ontological Terror,* 11.
18. Warren, 111.
19. Mitchell, *What Do Pictures Want?,* 237.
20. Cohen, "Denotation."
21. Mitchell, *What Do Pictures Want?,* 335.
22. Indigenous philosophies, while they no doubt differ from one another, tend not to divide and conquer life with dualisms. Enrolled member of the Citizen Potowatomi Nation and professor of environmental and forest

biology Robin Wall Kimmerer critiques the English pronoun "it," which is used for plants and other living beings: "It's no wonder that our language was forbidden. The language we speak is an affront to the ears of the colonist in every way, because it is a language that challenges the fundamental tenets of Western thinking—that humans alone are possessed of rights and all the rest of the living world exists for human use. Those whom my ancestors called relatives were renamed natural resources. In contrast to verb-based Potawatomi, the English language is made up primarily of nouns, somehow appropriate for a culture so obsessed with things." Kimmerer, "Speaking of Nature."

23. Pickering, *Mangle of Practice;* Latour, *Pandora's Hope.*

24. Burnett, *Sounding,* 674.

25. Burnett, 674.

26. Burnett, 9.

27. Burnett, 9.

28. In *Primate Visions,* for example, Haraway insists that the primates themselves were, to some degree, coauthors of the science and of the stories she tells.

29. Bryld and Lykke, *Cosmodolphins,* 25.

30. Haraway, *Staying with the Trouble,* 31.

31. Merchant, *Death of Nature,* 10, 30. See also Kolodny, *Lay of the Land.*

32. Irigaray, *Marine Lover,* 47.

33. Johnson and Braverman, "Blue Legalities," 5.

34. Johnson and Braverman, 7.

35. Neimanis, "Held in Suspense," 45–46.

36. Bonnie Mann counters Kant's disinterest, arguing that "natural beauty exhorts us to an interest in existence . . . not, however, primarily or even significantly *self-interest.*" Mann, *Women's Liberation,* 165.

37. *Oxford English Dictionary,* "sublime," 1.3b, 1.3a.

38. Zakiyyah Iman Jackson writes that the "Age of Discovery" was also the "age of slavery and conquest," explaining that the "demand for taxonomical and hierarchical races is foundational to the project of assimilating newly 'discovered' plants and nonhuman animals into a system." Jackson, *Becoming Human,* 25.

39. Verne, *Twenty Thousand Leagues under the Sea.*

40. Oreskes, "Scaling Up Our Vision," 391.

41. Haraway, "Situated Knowledges."

42. Chakrabarty, "Climate of History."

43. Gómez-Barris, *Extractive Zone,* 6.

44. Haraway, "Situated Knowledges."

45. Gómez-Barris, *Extractive Zone,* 100.

46. Chang, *World,* 127. Chang underscores "the political urgency of teaching geography in Hawai'i," analyzing a serialized geography of the world published in 1877, which features a "distinctly Hawaiian and ocean-centered perspective on the world, a perspective that countered the Western, colonialist, and land-centered perspective that dominated the textbooks used in schools in Hawaii" (103).

47. Ingersoll, *Waves of Knowing,* 22.

48. DeLoughrey, *Allegories,* 136.

49. The term "big science" was coined by Alvin M. Weinberg, an American nuclear physicist, in the 1960s; see Weinberg, *Reflections.* Niki Vermeulen explains that marine science has long been big in terms of extensive collaboration and expense: "Investigations into the oceans and their living creatures is big science *avant la lettre.* . . . In marine biology, large-scale collaboration is not only stimulated through the globally dispersed nature of the research material but also through a multi-disciplinary approach. . . . Moreover, technology is a reason to collaborate as costs are high." Vermeulen, "From Darwin to the Census of Marine Life."

50. Moreton-Robinson, *White Possessive,* 52.

51. "We are a circumpolar people, numbering approximately 160,000 Inuit spread across Chukotka (Russia), Alaska, Canada and Greenland. . . . Inuit own or have jurisdiction over half the Arctic; we are the largest Indigenous landholders in the world." "Inuit Nunangat," Indigenous Peoples Atlas of Canada, https://indigenouspeoplesatlasofcanada.ca/article/inuit-nunangat/.

52. Weaver, *Red Atlantic,* 260.

53. Teevee (Kinngait), "Sedna's Creation" and "Sedna's Wonder."

54. Teevee (Kinngait), "Untitled (Sedna by the Sea)."

55. Cooley, "Listening."

56. Iheka, *African Ecomedia,* 47.

57. Serres, *Malfeasance,* 73.

58. Serres, 73.

59. Tuck and Yang, "Decolonization Is Not a Metaphor."

60. The scientific team that Vescovo supported claims to have discovered "upwards of 40 new species," and the upbeat online story from BBC News quotes Patrick Lahey, the founder of the company that built the submarine, noting, "You're starting to see more privately funded marine research being conducted by wealthy individuals who bought subs they thought they would use recreationally but are now using to complete scientific expeditions." Amos, "Victor Vescovo." See also Young, *Expedition Deep Ocean.*

61. Broad, "Titan Disaster."

62. King, *Black Shoals,* 87.

63. King, 87.
64. See Nixon, *Slow Violence.*
65. See Shiva, *Biopiracy.*
66. Knowlton, *Citizens of the Sea.*
67. Latour, "Attempt," 484.
68. Latour, 488.
69. Stengers, "Challenge," 86.
70. Stengers, 100.
71. Scales, *Brilliant Abyss,* 204, 115.
72. Scales, xiii.
73. Scales, 199.
74. Radomska and Åsberg, "Fathoming Postnatural Oceans," 13.
75. "Indigenous Voices" petition.
76. See Alaimo, "Oceanic Origins," in *Exposed.*
77. Blue Climate Initiative, https://www.blueclimateinitiative.org/working-groups/mineral-and-genetic-resources. The Blue Climate Initiative is a "flagship program of the United Nations Decade of Ocean Science for Sustainable Development," https://www.blueclimateinitiative.org/sites/default/files/2023-10/BCI-Brochure.pdf.
78. von Uexküll, *Foray,* 75.
79. Marder, *Plant Thinking,* 62.
80. Balcombe, *What a Fish Knows,* 130.
81. Flusser and Bec, *Vampyroteuthis Infernalis,* 38.
82. Dickinson, "The brain—is wider than the sky."
83. Wolfe, *Ecological Poetics,* viii.
84. Wolfe, x.
85. Wolfe, x.
86. Wolfe, "Jagged Ontologies," 219.
87. Irigaray, *Marine Lover,* 47.
88. Graham, "Deep Water Trawling," 6.
89. Graham, 6, 7.
90. Graham, 6.
91. Graham, 7.
92. Deep Sea Conservation Coalition, "Game Over," http://www.savethehighseas.org.
93. Reid, "Solwara 1," 41.
94. Radomska and Åsberg, "Fathoming Postnatural Oceans," 3.
95. Haraway, *Staying with the Trouble,* 35.
96. Berger, *About Looking,* 26. The actual quote is, "Everywhere animals disappear. In zoos they constitute the living monument to their own disappearance."

97. Silverman, *World Spectators,* 128, 129.
98. Silverman, 92.
99. Silverman, 133.
100. Silverman, 133.
101. Silverman, 144.
102. Silverman, *Miracle of Analogy,* 11.
103. Silverman, 11.
104. Ledgard, *Submergence,* 288.
105. Nagel, "What Is It Like to Be a Bat?"
106. Mentz, *At the Bottom,* ix.
107. Puig de la Bellacasa, *Matters of Care,* 110.
108. Barad, *Meeting the Universe Halfway.*

1. Animating Surreal Creatures a Half Mile Down

1. Steinberg, *Social Construction.*
2. Mitman, *Reel Nature,* 58.
3. On hard bodies, see Jeffords, *Hard Bodies.* Readers interested in early African American marine biologists may be interested in Dr. Samuel Nabrit, who earned a PhD in marine biology at Brown in 1932 and who did research on fish at Woods Hole. See Nabrit, "Samuel Nabrit." Roger Arliner Young also worked at Woods Hole starting in 1927 and was the first African American woman to earn a PhD in zoology. See San Diego Supercomputer Center, "Roger Arlinger Young."
4. See Moreton-Robinson, *White Possessive.*
5. Jackson, *Becoming Human,* 23.
6. Voon, "Colonial Legacy."
7. Voon.
8. Elias, *Coral Empire,* 8.
9. The Wildlife Conservation Society (formerly the New York Conservation Society) posted an official apology for Grant and Osborne's white supremacy and for the despicable treatment of Ota Benga on their website: "Reckoning with Our Past, Present, and Future," July 29, 2020, https://www.wcs.org/reckoning-with-our-past-present-and-future-at-wcs. Gould, in *Remarkable Life of William Beebe, Explorer and Naturalist,* notes that Beebe was "constitutionally indifferent to politics," and that "although he deplored the inhumane treatment of native peoples by both European and other indigenous groups, his greatest concern was for the myriad animal species that were being hunted or marginalized to extinction." Beebe was concerned—selfishly, it seems—about the public relations implications of Grant's racism. Gould writes, regarding 1938 and 1939, "Will was not alone

in worrying that Grant's ugly racial politics, shared by many otherwise rational Americans at the time, would alienate people who had long been good friends to the zoo and to Will personally, and create irreparable divisions within the board and trustees" (176, 344).

10. Drayton, interview, 69.

11. Klein, "They Mixed Science."

12. McKittrick, *Dear Science,* 121.

13. See, for example, Quashie, *Black Aliveness.*

14. Elias, *Coral Empire,* 108.

15. Elias, 97–98.

16. Beebe, *Beneath Tropic Seas,* 5.

17. Cohen, "Denotation," 107.

18. Beebe, *Beneath Tropic Seas,* 6.

19. Official William Beebe website, https://sites.google.com/site/cwilliambeebe/Home, accessed January 14, 2020.

20. Ballard, *Eternal Darkness,* 13.

21. Adamowsky, *Mysterious Science,* 162.

22. Roorda, *Ocean Reader,* 465.

23. Kroll, *America's Ocean Wilderness,* 69, 68.

24. Burnett, *Sounding,* 270.

25. Gould, *Remarkable Life,* 411.

26. Beebe Papers, October 27, 1953, Box 18, Folder 3.

27. Warren, *Ontological Terror;* Gumbs, *M Archive,* 10. See the introduction and chapter 4 for more on Gumbs and Warren.

28. Kroll, "William Beebe," 40.

29. Kroll, 46.

30. Morgan, "Sounding the Deep."

31. Sheldon, *Into the Deep;* Enik, *Bathysphere Boys;* Leyssen, *Mission;* Rosenstock, *Otis and Will.*

32. Fox, *Bathysphere Book,* jacket blurb.

33. Dion, McLeod, and Thompson, *Exploratory Works.*

34. Dion, McLeod, and Thompson, 51.

35. Schlee, *History of Oceanography;* Idyll, *Abyss.*

36. Earle, *Sea Change,* 138.

37. Ballard, *Eternal Darkness,* 30.

38. Ballard, 30.

39. Broad, *Universe Below,* 39.

40. Koslow, *Silent Deep,* 54.

41. Berger, *Ocean,* 280.

42. Berger, 305n7.

43. Adamowsky, *Mysterious Science,* 171.

44. According to Elias, Frank Hurley and J. E. Williamson found themselves in a similar position. "As explorers their task was to expand science through knowledge, but as entertainers their ambition was to excite the public imagination with films and startling images." Elias, *Coral Empire,* 186.

45. Beebe, quoted in Kroll, *America's Ocean Wilderness,* 90.

46. Beebe, quoted in Kroll, 90.

47. Kroll, 86.

48. Kroll, 94.

49. Interestingly, Margot Norris links the critique of anthropocentrism with disciplinary mixing: "It is no historical accident that the critique of anthropocentrism shifts from natural science through philosophy into realms of literary and artistic expression. I would argue that the nature of biocentric discoveries, that in their destabilization of the very concept of form, triggered a confounding and softening of the generic boundaries that separate categories of thought within and among traditional disciplines or fields of knowledge." Norris, *Beasts,* 220–21.

50. Van Dover, *Deep-Ocean Journeys,* 7.

51. Cohen, "Denotation," 106.

52. See Latour, *We Have Never Been Modern.*

53. Whitehead, *Adventures,* 146.

54. Cohen, "Denotation."

55. Whitehead, *Adventures,* 267.

56. My analysis is indebted to the work of feminist epistemologists, feminist science studies, and environmentally oriented epistemologies, which I have discussed in detail elsewhere.

57. Braidotti, *Posthuman Knowledge,* 158.

58. Conniff, *Species Seekers,* 323.

59. Conniff, 324.

60. Beebe, *Half Mile Down,* 122.

61. Beebe, 125.

62. Beebe, 171.

63. Wolfe, *Ecological Poetics,* x.

64. Beebe, *Half Mile Down,* 137.

65. Beebe, 171.

66. Beebe, "World of Bermuda Fish," 38–39.

67. Beebe, "Melaphaes/Silver Eel/Gonostoma," in "Mid-Ocean," Beebe Papers, C0661, Box 12, Folder 10, 1929.

68. Beebe, *Half Mile Down,* xi.

69. Beebe, xv–xvi.

70. Firebrace, *Memo,* 77.

71. Daston and Galison, *Objectivity,* 372.

72. Daston and Galison, 374.
73. Kingsland, *Evolution of American Ecology,* 99.
74. Beebe, *Book of Naturalists,* 92.
75. Beebe, 87.
76. Beebe, 13.
77. Beebe, 96.
78. On the ecodelic, see Doyle, *Darwin's Pharmacy.*
79. Beebe, "Exploring a Tree and a Yard of Jungle," 113; Haraway, "Situated Knowledges"; Pollan, *How to Change Your Mind;* Doyle, *Darwin's Pharmacy;* Alaimo, "Your Shell on Acid," in *Exposed.*
80. Walls, *Passage to Cosmos,* 9.
81. Walls, 226.
82. Beebe, *Beneath Tropic Seas,* 86.
83. Stengers, *Thinking with Whitehead,* 42.
84. Beebe, *Beneath Tropic Seas,* 87, 90.
85. Beebe, "Mid-Ocean."
86. Steven Shaviro, *Without Criteria,* 63.
87. Shaviro, 62.
88. Beebe, "The Mystery of the Black Swallower," in "Mid-Ocean."
89. Bostelmann, the most distinctive artist on Beebe's team, also speculated about her piscine specimens, sometimes suggesting the cruelty of their capture. This passage, for example, narrates a story that positions the fish as agents in their own demise, "making" a journey and "finding" a destiny: "The fish have made a long journey up to my table and, far from their home in eternal night, they have found an unsought destiny." Imagining the brutal shifts in temperature and pressure, but gracefully obscuring culpability and causality, Bostelmann notes that "they had been drawn up into a sun-flooded world where they could not possibly adjust themselves," so for "those reasons they lay lifeless before me." Bostelmann, "Notes from an Undersea Studio," 129. We might hear echoes of Whitehead, quoting Wordsworth: "We murder to dissect." Whitehead, *Science and the Modern World,* 83.
90. Beebe, *"Lophodolus,"* Beebe Papers, Box 12, Folder 3, "Atlantic IInd Trip," September 24, 1929.
91. Beebe, *"Lophodolus."*
92. See Latour, *Pandora's Hope.*
93. Tee-Van, *"Lophodolus"*; Bostelmann, *"Lophodolus."*
94. *"Lophodolus acanthognathus* Regan, 1925."
95. Beebe, "Ctenophores," part of "Mid-Ocean."
96. Walls, *Passage to Cosmos,* 315.
97. Walls, 169.
98. Beebe Papers, Box 15, Folder 13.

99. Beebe, *Arcturus Adventure,* 201.

100. Beebe, 360.

101. Matsen lists Whitehead's *Science and the Modern World* as one of the books Beebe brought on his travels; Matsen, *Descent,* 17. Whitehead, *Science and the Modern World,* 87, 84.

102. Beebe, "Living Lamps," in "Mid-Ocean."

103. Beebe, "Living Lamps."

104. Shaviro, *Without Criteria,* 63–64.

105. Shaviro notes that Whitehead argues the world "is made of events, and nothing but events: happenings rather than things, verbs rather than nouns, processes, rather than substances" (17).

106. Beebe Papers, Box 12.

107. Current designations confirm *Argyopelecus* as a genus, within the family of Sternoptychidae. What Beebe refers to as a "silvery" hatchetfish is probably what is now designated as Sladen's hatchetfish, *Argyopelucus sladeni.* It is confusing, however, because "silver hatchetfish" is both a collective name for the genus and an alternative name for the species Sladen's hatchetfish. Moreover, *Argyropelecus hemigymnus,* the species designation Bostelmann uses for her painting of hatchetfish, is silvery as well. Taxonomy is a process. Currently there are seven extant species of *Argyropelecus.* See the World Registry of Marine Species and Search FishBase for more information.

108. Bostelmann, *"Argyropelecus hemigymnus."* This image appears in Beebe, "Round Trip," plate 3, "Silver Hatchetfish"; and in Beebe, "Half Mile Dive," 144.

109. Beebe Papers, Box 12, Folder 10.

110. Beebe, "Mid-Ocean."

111. Beebe, *Arcturus Adventure,* 298.

112. Beebe Papers, Box 12, Folder 8.

113. Beebe, *Half Mile Down,* 109.

114. Beebe, 109.

115. For more on elemental thinking, see the Elements book series at Duke University Press that I coedit with Nicole Starosielski and Courtney Berger.

116. Beebe, *Half Mile Down,* 111.

117. Osborn, "New Method," 28.

118. Beebe, *Half Mile Down,* 119.

119. Beebe, 132.

120. Beebe, 328.

121. Beebe, "Unknown Animals," Beebe Papers, Box 12, Folder 7.

122. Presumably "John" refers to DTR scientist John Tee-Van, who made two bathysphere descents, according to Matsen, *Descent.*

123. Beebe, *Half Mile Down,* 201.
124. Beebe, 203.
125. Pyle, *Art's Undoing,* 5.
126. Sehgal, "Aesthetic Concerns," 122.
127. Sehgal, 127.
128. Darby, "World Under Water," 688.
129. Mead, "Patterns of Culture," 686.
130. LaFollette, *Making Science Our Own,* 79.
131. Pound, "Retrospect," 12.
132. The scholarship on gender and modernism is too vast to discuss here. See Gilbert and Gubar, *No Man's Land,* for an early study.
133. Dion, McLeod, and Thompson, *Exploratory Works,* 51.
134. McLeod, "Sinking," 31.
135. Dion, McLeod, and Thompson, *Exploratory Works,* 52.
136. Dion, McLeod, and Thompson, 52.
137. Fessenden, "In the Early 20th Century."
138. McLeod, quoted in Fessenden. A letter from George Palmer Putnam, written to Beebe in 1925, is striking both for its misogynistic advice and for the fact that it anticipates Beebe's reaction: "You say you are going to ship off alone next time. I don't blame you. In fact I think a real he-man trip is exactly what you should do next. . . . My only amazement is at your fabulous courage—to go off on a floating wallowing steamer with so much femininity. Myself, I would kick seven of them in the teeth within fifty-six hours. . . . I think a stern order should go out from you, the commander, that the women are to be as little in evidence as possible at the landing. Certainly that none of them is to talk beyond the absolute necessities. (I can see you saying 'For God's sake! *This* coming from George.') The girls should say to the reporters 'Dr. Beebe will give out all the information.'" George Palmer Putnam, letter to William Beebe, May 20, 1925, Beebe Papers, Box 16, Folder 7: General Correspondence: Putnam, George,
139. "Ultraviolet Rays Make Rare Fish Transparent," 375.
140. WCS Archives, RG 9, 1900–71, Control 1005-a Box/Vol. 3; Beebe Papers, File Folder 1; Hollister, notebooks in Gloria Hollister Anabel Papers, New York Zoological Society, Department of Tropical Research, identifier 1006.
141. For more on Darwinian feminists see Alaimo, *Undomesticated Ground,* and Alaimo, "Sexual Matters."
142. Beebe, "Jellyfish and Equal Suffrage," 38.
143. Snitow, "Gender Diary," 9–43.
144. See Alaimo, *Undomesticated Ground.*
145. Beebe, "The Jellyfish," 47.

146. Beebe, 47.

147. The WCS Archives at the Bronx Zoo not only hold many of Bostelmann's original paintings and drawings but have also created a digital archive of the Department of Tropical Research, which includes her work: Wildlife Conservation Society Digital Archives, Department of Tropical Research Expedition Illustrations, 1916–53, https://library.wcs.org/Archives/Digital-Collections/DTR-Illustrations.aspx.

148. Conversation with Madeline Thompson, director of library and archives, at the Wildlife Conservation Society, Bronx Zoo, New York, summer 2017.

149. Roberts, "Surrealism and Natural History," 289.

150. Caws, *Surrealism,* 17.

151. Nadeau, *History of Surrealism,* 48.

152. Cahill, *Zoological Surrealism,* 25, 54.

153. Apollinaire, "Ocean of Earth," 179.

154. Moorhead writes that in 1936, Salvador Dalí donned a diving suit at the International Exhibition of Surrealism: "Dalí's idea was to give a lecture while wearing it. For good measure, he held two dogs on leads in one hand, and had a billiard cue in the other. During the course of the lecture, however, it became apparent that he was slowly suffocating inside his helmet, and it had to be prised off with the billiard cue. Dalí recovered and went on to finish his presentation with a slide show. The slides, it almost goes without saying, were presented upside down." Moorhead, "Dalí in a Diving Helmet."

155. Krause, "Scales Too Large." Krause draws on McLeod's knowledge about Bostelmann.

156. Schaffner, *Salvador Dalí's Dream of Venus.* Although it is women's bare breasts that are featured prominently in the photos, the organizers wanted the title to be "Bottoms of the Sea" (72, 74).

157. Styrsky, "Cuttlefish Man."

158. The WCS's digital collection of Bostelmann's paintings is organized by date, showing that the black background appears in 1930, although it could have appeared earlier, in 1929, as some paintings with black backgrounds are identified with a range of possible dates, starting with 1929.

159. Bostelmann obituary, *New York Times,* December 29, 1961.

160. Bostelmann's grandson, Michael Crumpton, in comments to the video "The Bathysphere: Discovering the Deep," says that Bostelmann told him "she had gone down in the Bathysphere to paint the amazing denizens," but I have found no documentation of this.

161. Adamowsky, *Mysterious Science,* 166.

162. Bostelmann, "Notes from an Undersea Studio," 129.

163. WCS Expedition, 1916–53.

164. Widder, "Fine Art," 171.
165. Widder, 174.
166. Widder, 171.
167. Widder, 177.
168. Dion, McLeod, and Thompson, *Exploratory Works,* 49.
169. Bostelmann, "Tangled Web." The WCS Archives include three illustrations by Bostelmann on black backgrounds dated 1930: "Spiny Larva Pursuing Copepods," *"Stylophthalmus,"* and the gorgeously colored and detailed "Omosudis Pursuing Red Squid." However, the vast majority of her paintings from 1930 and 1931 are on white backgrounds.
170. See, e.g., Bostelmann, "*Photosomtias guerni.*"
171. Bostelmann, "Dragon."
172. Bostelmann, untitled painting in Beebe.
173. Bostelmann, *"Bathysphaera intacta."*
174. Latour, "Why Has Critique Run Out of Steam?"
175. Bostelmann, *"Stylophthalmus."*
176. Gerspacher, *Owls,* 29.
177. Bostelmann, "Saber-Toothed Viper Fish."
178. Bostelmann, "Orange-Lighted."
179. Bostelmann, untitled work in Widder, "Fine Art."
180. Beebe Papers, Folder 22, File 28, 326.
181. Bostelmann, "Silver Hatchetfish."
182. Beebe, "Memorabilia," Beebe Papers, Box 18, Folder 6.
183. Bostelmann, "Silver Hatchetfish."
184. Gould, *Remarkable Life,* 293.
185. Gould, 293.
186. Gould, 294.
187. Fisher, *Weird,* 15.
188. Bostelmann, "Deep Sea Angler Fish."
189. Beebe, "Abyssal Life," Beebe Papers, Box 14, Folder 1, 9, 9–10.
190. Bostelmann, "Front View." This image was captioned "Big Bad Wolves of an Abyssal Chamber of Horrors" in a 1931 *National Geographic* article.
191. Cahill, *Zoological Surrealism,*119.
192. de Waal, "Anthropomorphism and Anthropodenial."
193. Wolfe, *What Is Posthumanism?,* 47.
194. Burnett, *Sounding,* 312.
195. Burnett, 316.
196. Burnett, 316–17.
197. Bostelmann, "Notes from an Undersea," 128.
198. Ray Troll and the Ratfish Wranglers, "Ballad of William Beebe."

2. Packed Up in Tupperware and Russian Vodka

1. "Bathysphere Used in War Research," 25.
2. Ballard, *Eternal Darkness,* 32.
3. Koslow, *Silent Deep,* 55.
4. Oreskes, *Science on a Mission,* 489.
5. Oreskes, 257.
6. Oreskes, 488.
7. Oreskes, 488–90.
8. Oreskes, 490.
9. Oreskes, 497.
10. Oreskes, 497.
11. Sehgal, "Aesthetic Concerns," 122.
12. Carson, *Under the Sea Wind,* 5.
13. Carson, 3.
14. Seager, "Radical Observation," 272.
15. Rachel Carson, *Under the Sea Wind,* 5.
16. Carson, 5.
17. Carson, 139.
18. Carson, 151.
19. Carson, 152.
20. Carson, 153.
21. Carson, 154, 155.
22. Carson, 154, 155.
23. Carson, 154.
24. Carson, 160.
25. Carson, 162.
26. Carson, 53, 49.
27. Carson, *Sea Around Us,* 45.
28. Scholarship on Wyndham often cites Brian Aldiss's take on *The Kraken* and *Day of the Triffids:* "Both novels were totally devoid of ideas but read smoothly, and thus reached a maximum audience who enjoyed cosy disasters." Aldiss, *Trillion Year Spree,* 254. See also Christopher Daley, "The Not So Cozy Catastrophe."
29. Cohen, referencing Jonathan Lamb, explains that "throughout published observations of the undersea, we find the phrase, 'words cannot express.' The rhetorical term, the *je ne sais quoi,* dates to the early modern era to underline the writer's failure to convey an intense experience—that, in twentieth century dive narratives, was the perception of extraordinary submarine reality." Cohen, *Underwater Eye,* 7.

30. On the logic of the supplement, see Derrida, *Of Grammatology;* for *différance,* see Derrida, *Margins.*

31. Vitale, *Biodeconstruction,* 22.

32. Helmreich and Roosth, "Life Forms," 34.

33. On the absurdity and violence of the very concept of the animal, see Derrida, *Animal.*

34. Helmreich, *Alien Ocean,* 17.

35. Wolfe, *Ecological Poetics,* xiv.

36. Wyndham, *Kraken Wakes,* 7.

37. Tennyson, "Kraken."

38. Wyndham, *Kraken Wakes,* 11.

39. Wyndham, 12, 14.

40. Wyndham, 16.

41. Wyndham, 26–29.

42. Wyndham, 31.

43. Wyndham, 31.

44. Wyndham, 33.

45. Wyndham, 34.

46. Wyndham, 34.

47. Wyndham, 44.

48. Wyndham, 45.

49. Wyndham, 46, 47.

50. Wyndham, 68–69.

51. Wyndham, 71

52. Wyndham, 84, 95, 107.

53. Ketterer, "John Wyndham," 382.

54. Ketterer, 382.

55. Ketterer, 382.

56. Ketterer, 383.

57. See Thomas, "Moments of Productive Bafflement."

58. Wyndham, *Kraken Wakes,* 125.

59. Wyndham, 138, 139.

60. Wyndham, 139–40.

61. Wyndham, 140–41.

62. Wyndham, 145, 146.

63. Wyndham, 154.

64. Wyndham, 182.

65. Wyndham, 240.

66. Wyndham, 230.

67. Wyndham, 61.

68. Wyndham, 230.

69. Wyndham, 239.
70. Wyndham, 239.
71. Wyndham, 240.
72. Vint, *Animal Alterity,* 11.
73. Vint, 157.
74. Csicsery-Ronay, *Seven Beauties,* 47, 51.
75. Csicsery-Ronay, 55
76. Vint, *Animal Alterity,* 227.
77. Clarke, *Deep Range,* 245.
78. Clarke, *Coast of Coral,* 3, 264.
79. Cohen, *Underwater Eye,* 90.
80. Cohen, 93.
81. Cohen, 93.
82. Diolé, *Undersea Adventure,* 165–66.
83. Diolé, *Undersea Adventure,* plate 9.
84. Diolé, plates 13a, 14, and 15.
85. Clarke, *Coast of Coral,* 50
86. Burnett writes that Gifford B. Pinchot (son of the well-known Gifford Pinchot) and his colleagues in the early 1960s "laid out a plan to convert a set of Pacific atolls into whale farms where captive rorquals (towed in floating cages into the circular bays as juveniles) would be fattened on artificial plankton blooms until they were large enough to slaughter for meat, oil, and other by-products" (Burnett, *Sounding,* 523). Just as the novel's protagonist imagines himself as the whales' protector, Pinchot thought his plan would save whales from extinction—as Burnett puts it, "domestication being a superb bulwark against extermination" (524). Perhaps the whale farming scenario Clarke describes in the novel inspired Pinchot: the short story was published in 1953 and the novel was first published in 1957. Ballard attests to plans for extracting riches from the depths: "By the mid-1960s, an assorted group of scientists, analysts and futurists were selling the deep sea as America's next manifest destiny, waiting to reward the first to arrive in the dark abyss with untold riches—new commercial fisheries, oil reserves, precious gems and metals" (Ballard, *Eternal Darkness,* 81).
87. Clarke, *Deep Range,* 420.
88. Clarke, 456.
89. Clarke, 442, 443.
90. Clarke, 487.
91. Clarke, 421.
92. Clarke, 361.
93. Clarke, 263.
94. Clarke, 357, 358.

95. Clarke, 360, 361.
96. Clarke, 362.
97. Clarke, 366.
98. Clarke, 370, 371.
99. Clarke, 374.
100. Clarke, 376.
101. Clarke, 380.
102. Clarke, 388.
103. Clarke, 394.
104. Clarke, 394.
105. Clarke, 395.
106. Berger, *Ways of Seeing,* 26.
107. Kaharl, *Water Baby,* 96.
108. Kaharl, 96.
109. Kaharl, 96.
110. Kaharl, 177.
111. Ballard, *Eternal Darkness,* 171.
112. See Shukin, *Animal Capital,* for more on rendering.
113. Hamblin, *Oceanographers,* xxiii.
114. "Dive to the Edge of Creation."
115. Ballard, *Eternal Darkness,* 109.
116. Kaharl, *Water Baby,* 165.
117. Kaharl, 91.
118. Ballard, *Eternal Darkness,* 170, 171, 173.
119. Ballard, 173.
120. A recent international study identifies the experience of awe with the mental health and "flourishing" of scientists. See Jacobi et al., "Aesthetic Experiences."
121. Kaharl, *Water Baby,* 97.
122. Kaharl, 98.
123. Kaharl, 236.
124. Kaharl, 236, 233.
125. Oreskes, *Science on a Mission,* 340, 339.
126. Oreskes, 328.
127. Oreskes, 374.
128. Oreskes, 374, 376.
129. Oreskes, 383.
130. Kaharl, *Water Baby,* 345.
131. Kaharl, 101. Readers may be interested to know that Evan Forde in 1979 "became the first Black scientist to participate in research dives aboard a deep-sea submersible." In 1980 he descended in *Alvin.* See Florida State

University Coastal and Marine Laboratory, "Black Pioneers of the Marine Sciences," February 10, 2021, updated February 26, 2024, https://marinelab.fsu.edu/news-at-fsucml/black-history-month-2024/; and "Evan Ford Describes Being on a Submersible Dive," video and written transcript, tape 6, History Makers, June 3, 2013, https://da.thehistorymakers.org/story/27124.

132. Van Dover, "Deep in the Sea."

133. Van Dover, *Deep-Ocean Journeys,* 2.

134. Van Dover, 77.

135. Van Dover, 77.

136. Van Dover, 79.

137. Van Dover, 79.

138. Van Dover, 169, 171.

139. Van Dover, "Hydrothermal Vent Ecosystems," 315.

140. Watts's *Starfish* was originally conceived as a stand-alone novel but was then followed by three more volumes; the trilogy is set in 2050–56. See Watts, "Wildlife," 606.

141. Watts, *Starfish,* 9.

142. Watts, *Starfish,* 11.

143. Watts, 34.

144. Watts, 61.

145. Wall, "Here Be Monsters," 79.

146. Watts, *Starfish,* 74.

147. Watts, 74.

148. Watts, "Wildlife," 613.

149. Watts, *Starfish,* 113.

150. Watts, 113.

151. Broad, *Universe Below,* 107, 134.

152. Broad, 109.

153. Helmreich, *Alien Ocean,* 284.

154. Helmreich, 284.

155. Watts, *Starfish,* 88.

156. The bio on Watts's website (https://www.rifters.com/real/author.htm), introduced by the epigraph "Whenever I find my will to live becoming too strong, I read Peter Watts," is revealing: "Peter Watts is an awkward hybrid of biologist, science-fiction author, and (according to the U.S. Department of Homeland Security) convicted felon/tewwowist *[sic].* In addition to a number of accolades for science fiction, he has won minor awards in fields as diverse as marine mammal research and video documentary. None of these have gone to his head since they never involved a lot of cash. He spent ten years getting a bunch of degrees in the ecophysiology of marine mammals and another ten trying to make a living on those qualifications without

becoming a whore for special-interest groups. This proved somewhat tougher than it looked; throughout the nineties he was paid by the animal welfare movement to defend marine mammals; by the U.S. fishing industry to sell them out; and by the Canadian government to ignore them."

157. Watts, *Starfish,* 127.

158. Watts, 189.

3. Counting and Framing

1. Escobar, *Designs for the Pluriverse,* 222; Watkins, *Gold Fame Citrus,* 114.

2. See Alaimo, "Your Shell on Acid," in *Exposed.*

3. See Oreskes, *Science on a Mission.*

4. Helmreich, *Alien Ocean,* xi.

5. Helmreich, 284.

6. King, *Black Shoals,* 87.

7. "Mariana Trench First-Ever Look."

8. Jamieson, *Hadal Zone,* 278.

9. Jamieson, 278.

10. Jamieson, "Meet the Mysterious 'Monsters.'"

11. Haraway, *Primate Visions.*

12. Haraway, 153.

13. Ingram, *Imperiled Whiteness,* 13.

14. Gumbs, *Undrowned,* 118.

15. Helmreich, *Alien Ocean,* 175.

16. Helmreich, 175.

17. Helmreich, 196.

18. *Wired,* cover August 2004.

19. See Alaimo, *Bodily Natures,* chap. 6, for more critiques of genetic discourses, including those of Richard Lewontin and Richard Levins, Evelyn Fox Keller, Bonnie Spanier, and Donna J. Haraway.

20. Shreeve, "Craig Venter's Epic Voyage," 111.

21. Pickering, *Mangle of Practice.*

22. Conover, *Cracking the Ocean Code,* 2007.

23. Shreve, "Craig Venter's Epic Voyage," 150.

24. Wolfe, *Before the Law,* 103, 104.

25. Wolfe, 103.

26. Here I am referencing Beck, *Risk Society,* along with my conception of transcorporeality in Alaimo, *Bodily Natures.*

27. Broglio, *Animal Revolution.*

28. Cressey, "Out of the Blue," 514.

29. Census of Marine Life website, http://www.coml.org.

30. Ausubel, "What Lies Beneath," 19.

31. Ocean Biodiversity Information System (OBIS) website, https://obis.org, accessed 2020.

32. Baker, Ramirez-Llodra, and Tyler, introduction to *Natural Capital,* 17.

33. Bouchet, "Magnitude," 41.

34. Bouchet, 33.

35. Grassle and Maciolek, "Deep-Sea Species Richness," 315.

36. Grassle and Maciolek, quoting Nybakken, 315.

37. Grassle and Maciolek, 318, 336.

38. Bouchet, "Magnitude," 58.

39. Oreskes, *Science on a Mission,* 497.

40. Rex and Etter, *Deep-Sea Biodiversity,* 243.

41. Rex and Etter, x.

42. Snelgrove and Smith, "Riot of Species," 311–42.

43. Jamieson, *Hadal Zone,* 282.

44. Youatt, *Counting Species,* 1.

45. Youatt, 28.

46. Sehgal, "Aesthetic Concerns," 122.

47. Haraway, "Situated Knowledges," 189.

48. Ausubel, Crist, and Waggoner, *First Census,* 35.

49. Snelgrove, *Discoveries,* xi.

50. Snelgrove, 178.

51. Hau'ofa, "Our Sea of Islands."

52. Snelgrove, *Discoveries,* ix–x.

53. Haddock, quoted in Crist, Scowcroft, and Harding, *World Ocean Census,* 175.

54. Haddock, 194.

55. Daston and Galison, *Objectivity,* 369.

56. Beebe, *Half Mile Down,* xi.

57. Daston and Gallison, *Objectivity,* 272.

58. This is not to argue, of course, that conventional science is not essential for diagnosing, understanding, and creating solutions for many anthropogenic crises of the twenty-first century. Given climate deniers, Covid-19 deniers, and antivaxxers, knee-jerk reactions against science, medicine and expertise of all sorts have been articulated with the political right. Yet critiques of science from the left remain important.

59. Ausubel, "Foreword," vii–viii.

60. Ausubel, vii.

61. Ausubel, vii–viii.

62. Ebbe et al., "Diversity," 154–55.
63. Ebbe et al., 157.
64. Knowlton, *Citizens of the Sea,* 9.
65. Knowlton, 9.
66. Horton, *Cosmic Zoom,* 3.
67. Berland, *Virtual Menageries,* 4.
68. Ausbel, "What Lies Beneath," 21.
69. Nouvian, *The Deep.*
70. Heise, *Imagining Extinction,* 65.
71. Census of Marine Life website, "Image Gallery," http://www.coml.org/image-gallery. This website no longer works.
72. Jue, *Wild Blue Media,* 121.
73. Jue, 121.
74. Census of Marine Life, "Image Gallery."
75. Wolfe, *Before the Law,* 9.
76. Wolfe, 6.
77. Latour, *On the Modern Cult,* 93.
78. Latour, 71.
79. Census of Marine Life, "Image Gallery," "New Species."
80. Census of Marine Life, "Image Gallery."
81. Latour, *On the Modern Cult,* 123.
82. Latour, *Pandora's Hope.*
83. Berland, *Virtual Menageries,*120.
84. "Hard to See Creatures."
85. "Hard to See Creatures."
86. Latour, *On the Modern Cult,* 84.
87. "Deep Sea Creatures," video.
88. Jamieson, *Hadal Zone,* 125.
89. Shapin and Schaffer, *Leviathan and the Air-Pump,* 23.
90. Census of Marine Life website.
91. Nouvian, *The Deep,* 12.
92. Nouvian, *The Deep,* 12.
93. Nouvian, "Meet Claire Nouvian."
94. Nouvian, *The Deep,* 252.
95. Rex and Etter, *Deep-Sea Biodiversity,* 10.
96. Rex and Etter, 10. The lack of ecological information about deep-sea life in the McGraw-Hill textbook published in 2010 is visually apparent because compared to the other chapters included in part 3, on the topic of marine ecosystems, there are only two small photos that depict more than one organism; nearly all the illustrations depict individual species. Castro and Huber, *Marine Biology.*

97. Rex and Etter, figures A5–A31.

98. Koslow, *Silent Deep,* plates 1–16.

99. Heise makes a similar argument about Joel Sartore's photographs, which "give no sense of habitat or ecological connectedness, a strategy that may seem at odds with environmentalist emphasis on locality and rootedness . . . [but thereby shifts] the conservationist appeal to a mostly aesthetic level." Heise, *Imagining Extinction,* 82–83.

100. Broglio, *Surface Encounters,* xxxi.

101. Email communication with Claire Nouvian, 2012.

102. Nouvian, *The Deep,* 26.

103. Nouvian, 40–42.

104. Nouvian, 11.

105. Nouvian, 12.

106. Nouvian, 4–5, 8–9.

107. Nouvian, 13, 14.

108. Nouvian, 34.

109. Nouvian, 50, 52–53.

110. Nouvian, cover image, 54–55.

111. Nouvian, 54.

112. Nouvian, 37.

113. Nouvian, 56, 65, 95, 118.

114. *The Deep,* exhibition.

115. *The Deep,* exhibition.

116. "Expedition: 'The Deep.' "

117. Broglio, *Surface Encounters,* 124.

118. South Korean filmmaker Bong Joon Ho is said to be making a film inspired by the creaturely palette in Nouvian's book. See Marc, "Director Bong Joon Ho."

119. Nouvian, *The Deep,* 11.

4. Clickbait, Black Void, or Intimate Mediation?

1. *Blue Peril.*

2. Hau'ofa, "Our Sea of Islands," 39.

3. *Blue Peril.*

4. *Blue Peril.*

5. *Blue Peril.*

6. *Blue Peril.*

7. This analysis is based on having been in the audience for one of McKibben's early 350.org events at University of Texas at Arlington, in which the content and the images, from around the globe, consistently

posed people without including other species in the frame. See the conclusion to Alaimo, *Exposed.*

8. *Oxford English Dictionary,* "Lure," 1 transitive, 2a intransitive.
9. Daston and Galison, *Objectivity,* 372.
10. Connors, personal website, https://www.aconnors.com/#:~:text=He%20started%20his%20career%20as,in%20the%20Google%20Research%20team.
11. Connors, *Girl,* 162, 163.
12. Connors, 372, 383.
13. Connors, 6, 60.
14. Connors, 61.
15. McVeigh, "More than 5,000 New Species."
16. Connors, *Girl,* 370.
17. Kraese, *Deep-Sea Creeps,* 3.
18. Kraese, 4.
19. *Deep Sea (Shen hai).*
20. *Deep Abyss.*
21. Ledgard, *Submergence,* 167.
22. Ledgard, 167.
23. Ledgard, 174.
24. Darwin, *Origin of Species,* 528, 529. On Darwin, see Alaimo, "Sexual Matters," and Alaimo, "Oceanic Origins, Plastic Activism, and New Materialism at Sea," in *Exposed.*
25. Ledgard, *Submergence,* 196.
26. Ledgard, 196, 197.
27. Ledgard, 206.
28. Ledgard, 208.
29. Ledgard, 209.
30. Baudrillard, *Simulacra and Simulation.*
31. *Meg 2.* Thanks to Michael Allan for recommending this film—a perfect fit for this book!
32. Koslow, *Silent Deep,* 18. For more on living fossils in the deep, see Alaimo, "Spectral Species."
33. Barclay and Tidwell, "Introduction," 269, 271.
34. "The Deep," *Blue Planet II.*
35. Warren, *Ontological Terror,* 110.
36. Warren, 111.
37. Warren, 1.
38. Jameson et al., "Fear and Loathing," 803.
39. Womack notes the transatlantic significance of Mami Wata, which links Okorafor's novel to the discussion of Gumbs that follows: "The Mami

Wati are also closely associated with Africans brought to the New World in the transatlantic slave trade. They inspired the Drexciya myth, of female slaves thrown overboard who now live in the sea." Womack, *Afrofuturism,* 87. See also Iheka, *African Ecomedia,* for contemporary African art that references Mami Wata.

40. Jue, "Intimate Objectivity," 174.
41. Okorafor, *Lagoon,* 3.
42. Okorafor, 3.
43. Okorafor, 5.
44. Okorafor, 5.
45. Okorafor, 6.
46. Butler, *Lilith's Brood.*
47. Okorafor, *Lagoon,* 91.
48. Okorafor, 111.
49. Okorafor, 245.
50. Wilkes, "Poison Rainbows," 307.
51. Okorafor, *Lagoon,* 278.
52. Ikheka, *African Ecomedia,* 27.
53. Gumbs, *M Archive,* 7.
54. Gumbs, 7.
55. Gumbs, 10.
56. Gumbs, 10.
57. Gumbs, 10, no capitalization in original.
58. Gumbs, 11.
59. Gumbs, *Dub,* xii.
60. Kornbluh, *Immediacy,* 15.
61. Kornbluh, 17.
62. Kornbluh, 153.
63. Kornbluh includes the "fascination" with "imagined Indigenous epistemologies" in her critique, with no explanation for that dismissal (174).
64. Kornbluh, 6.
65. Jue, *Wild Blue Media,* 3.
66. Pickering, *Mangle of Practice.* Latour, *Pandora's Hope.*
67. Berger, *About Looking,* 19, 22.
68. Hayward, "Sensational Jellyfish," 164, 165.
69. Hayward, 165.
70. *Into the Deep,* exhibit.
71. Alan Jamieson and coauthors have documented "extraordinary levels of persistent organic pollutants in the endemic amphipod fauna from two of the deepest ocean trenches." Jamieson et al., "Bioaccumulation," 1. Jamieson and Ondo, "Lebensspuren and Müllspuren," have also documented benthic

trails made by plastic bags moving across the seafloor. Indeed, Weston et al. have named a newly discovered species of deep-sea amphipod dwelling in the Mariana Trench *Eurythenes plasticus,* for the microplastic fiber found in its hindgut—evidence that plastic pollution extends to the deep-sea floor. Weston et al., "New Species of *Eurythenes.*"

72. Haraway, *Companion Species Manifesto.*
73. Kornbluh, *Immediacy,* 151.
74. Montemurno, "Meet the Giant Isopod."
75. Fay, *Inhospitable World,* 201.
76. Fay, 201.
77. Escobar, *Designs,* 222.

Epilogue

1. Jamieson et al., "Fear and Loathing," 797, 798.
2. Jamieson et al., 798.
3. Jamieson et al., 805.
4. Armstrong et al., "People Do Care."
5. Jamieson et al., "Reply to People Do Care," 3.
6. Beauman, *Venomous,* 7.
7. "Meet the Lumpsucker."
8. Beauman, *Venomous,* 18.
9. Beauman, 129.
10. Beauman, 48.
11. Beauman, 77.
12. Beauman, 81.
13. Beauman, 115.
14. Beauman, 186.
15. Vecchione et al., *Deep Ocean,* 94.
16. Mann, *Women's Liberation,* 163.

Bibliography

The Abyss. Directed by James Cameron. Film. 20th Century Studios, 1989.

Adamowsky, Natascha. *The Mysterious Science of the Sea, 1775–1943*. London: Routledge, 2015.

Alaimo, Stacy. *Bodily Natures: Science, Environment, and the Material Self*. Bloomington: Indiana University Press, 2010.

Alaimo, Stacy. *Exposed: Environmental Politics and Pleasures in Posthuman Times*. Minneapolis: University of Minnesota, 2016.

Alaimo, Stacy. "Sexual Matters: Darwinian Feminism and the Nonhuman Turn." *J19: Journal of Nineteenth Century Americanists* 1, no. 2 (2013): 390–96.

Alaimo, Stacy. "Spectral Species in the (Political) Abyss: From Living Fossils to Presumed Extinctions." In *Blue Extinctions*, edited by Rachel Murray and Veronica Fibisan. London: Palgrave Macmillan, 2025 (forthcoming).

Alaimo, Stacy. *Undomesticated Ground: Recasting Nature as Feminist Space*. Ithaca, N.Y.: Cornell University Press, 2000.

Aldiss, Brian Wilson, with David Wingrove. *Trillion Year Spree: The History of Science Fiction*. London: Gollancz, 1986.

Aliens of the Deep. Directed by James Cameron and Steven Quale. Film. Buena Vista Studios, 2005.

Amos, Jonathan. "Victor Vescovo: Adventurer Reaches Deepest Ocean Locations." BBC News, September 9, 2019. https://www.bbc.com/news/science-environment-49636756.

Apollinaire, Guillaume. "Ocean of Earth." 1918. Translated by Roger Shattuck. In *The Sources of Surrealism: Art in Context*, edited by Neil Matheson, 179. London: Lund Humphries, 2006.

Armstrong, Claire W., Margrethe Aanesen, Stephen Hines, and Rob Tynch. "People Do Care about the Deep Sea. A Comment on Jamieson et al." *ICES Journal of Marine Science* 79 (2022): 2336–39.

Ausubel, Jesse H. Foreword to *Life in the World's Oceans: Diversity, Distribution, and Abundance*, edited by Alasdair D. McIntyre, vii–viii. Oxford: Wiley-Blackwell, 2010.

Ausubel, Jesse. “What Lies Beneath: An Interview with Jesse Ausubel, Cofounder, Census of Marine Life.” Conducted by Amy Entwisle. *Imagine,* January/February 2011, 18–21.

Ausubel, Jesse H., Darlene Trew Crist, and Paul E. Waggoner. *First Census of Marine Life, 2010: Highlights of a Decade of Discovery.* New York: Census of Marine Life, 2010. http://www.coml.org/comlfiles/partner2010/Final%20Report%209-14%20small.pdf.

Baker, Maria, Eva Ramirez-Llodra, and Paul Tyler. *Natural Capital and Exploitation of the Deep Ocean.* Oxford: Oxford University Press, 2020.

Balcombe, Jonathan. *What a Fish Knows: The Inner Lives of Our Underwater Cousins.* New York: Scientific American, 2016.

Ballard, Robert D. *The Eternal Darkness: A Personal History of Deep-Sea Exploration.* Princeton, N.J.: Princeton University Press, 2000.

Barad, Karen. *Meeting the Universe Halfway: Quantum Physics and the Entanglement of Matter and Meaning.* Durham, N.C.: Duke University Press, 2007.

Barclay, Bridgitte, and Christy Tidwell. “Introduction: Mutant Bears, Defrosted Parasites and Cellphone Swarms: Creature Features and the Environment.” *Science Fiction Film and Television* 14, no. 3 (2021): 269–77.

“Bathysphere Used in War Research.” *New York Times,* January 12, 1944.

Baudrillard, Jean. *Simulacra and Simulation.* Ann Arbor: University of Michigan Press, 1994.

Beauman, Ned. *Venomous Lumpsucker.* New York: Soho, 2022.

(Beebe Papers) William Beebe Papers, Department of Rare Books and Special Collections, Princeton University Library. https://library.princeton.edu/special-collections/collections/william-beebe-papers.

Beebe, William. *The Arcturus Adventure.* New York: Putnam, 1926.

Beebe, William. *Beneath Tropic Seas: A Record of Diving among the Coral Reefs of Haiti.* New York: Putnam's Sons, 1928.

Beebe, William. “The Depths of the Sea: Strange Life Forms a Mile below the Surface.” *National Geographic* 8 (January 1, 1932): 65–88.

Beebe, William. “Exploring a Tree and a Yard of Jungle.” In Dion, McLeod, and Thompson, *Exploratory Works,* 106–14.

Beebe, William. “A Half Mile Dive in the Bathysphere.” *Bulletin: New York Zoological Society* 35, no. 5 (1932): 143–46.

Beebe, William. *Half Mile Down.* Chicago: Cadmus, 1934.

Beebe, William. “The Jellyfish and Equal Suffrage.” *Atlantic,* July 1914, 36–47.

Beebe, William. “A Round Trip to Davy Jones's Locker: Peering into Mysteries a Quarter Mile Down in the Open Sea, by Means of the Bathysphere.” *National Geographic* 59, no. 6 (1931): 653–78.

Beebe, William. "The World of Bermuda Fish." *New York Zoological Society, Department of Tropical Research, Bulletin* 35, no. 2 (1932).

Beebe, William, ed. *The Book of Naturalists: An Anthology of the Best Natural History.* Princeton, N.J.: Princeton University Press, 1944.

Beck, Ulrich. *Risk Society.* Translated by Mark Ritter. Los Angeles: Sage, 1992.

Berger, John. *About Looking.* New York: Vintage, 1980.

Berger, John. *Ways of Seeing.* London: Penguin, 1972.

Berger, Wolf H. *Ocean: Reflections on a Century of Exploration.* Berkeley: University of California Press, 2009.

Berland, Jody. *Virtual Menageries: Animals as Mediators in Network Culture.* Cambridge, Mass.: MIT Press, 2019.

Blue Peril. Film. Deep Sea Mining Campaign (DSMC), 2022. https://dsm-campaign.org/blue-peril/.

Bostelmann, Else. "*Argyropelecus hemigymnus:* Silver Hatchetfish Drifting through Abyssal Darkness." WCS Archives, WCS-1039-01-05-U077-R-CR, 1929–31.

Bostelmann, Else. "*Bathysphaera intacta* Circling the Bathysphere." 1934. WCS Archives, WCS-1039-01-05-1002-R-CR.

Bostelmann, Else. "Deep Sea Angler Fish." Between 1929 and 1937. WCS Archives, WCS-1039-01-05-U046-R-CR.

Bostelmann, Else. "The Dragon of the Shining Green Bow." In Beebe, "Depths of the Sea," 65–88.

Bostelmann, Else. "Else Bostelmann, Artist, Dies at 79: Painter of Sea Life Served on Beebe Expeditions." Obituary. *New York Times,* December 29, 1961. https://www.nytimes.com/1961/12/29/archives/else-bostel-mann-artist-dies-at-79-painter-of-sea-life-served-on.html.

Bostelmann, Else. "Front View of 8 Deep Sea Fish of Which 5 Are Seen from the Bathysphere," 1931. WCS Archives, WCS-1039-01-1033-R-CR.

Bostelmann, Else. *"Lophodolus."* 1930. WCS Archives, WCS-1039-01-05-0617-R-CR.

Bostelmann, Else. "Notes from an Undersea Studio off Bermuda." In Dion, McLeod, and Thompson, *Exploratory Works,* 126–30.

Bostelmann, Else. "Orange-Lighted Finger-Squid Catching Lanternfish" (caption). In Beebe, "Round Trip."

Bostelmann, Else. "*Photosomtias guerni* and *Acanthyphyra purpura.*" Circa 1939, WCS Archives, WCS-1039-01-05-U084-R-CR.

Bostelmann, Else. "Saber-Toothed Viper Fish *(Chauliodus sloanei)* Chasing Ocean Sunfish *(Mola mola).*" 1934. WCS Archives, WCS-1039-01-05-1001-R-CR.

Bostelmann, Else. *"Stylophthalmus."* WCS Archives, WCS-1039-01-05-U081-R-CR.

Bostelmann, Else. "The Tangled Web of Death of a Deep Sea Squid." Painting in Beebe, "Depths of the Sea," 78.

Bostelmann, Else. Untitled painting of bathysphere and parrotfish. In Beebe, "Half Mile Dive," 180.

Bostelmann, Else. Untitled work. In Widder, "Fine Art," fig. 2, p. 173.

Bouchet, Phillip. "The Magnitude of Marine Biodiversity." In *The Exploration of Marine Biodiversity: Scientific and Technological Challenges,* edited by Carlos M. Duarte, 33–63. Paris: Fundación BBVA, 2006.

Braidotti, Rosi. *Posthuman Knowledge.* Cambridge: Polity, 2019.

Broad, William J. *The Universe Below: Discovering the Secrets of the Deep Sea.* New York: Touchstone, 1997.

Broad, William J. "Titan Disaster Forces Global Rethinking of Deep Sea Exploration." *New York Times,* June 18, 2024. https://www.nytimes.com/2024/06/18/science/titan-submersible-anniversary.html.

Broglio, Ron. *Animal Revolution.* Minneapolis: University of Minnesota Press, 2022.

Broglio, Ron. *Surface Encounters: Thinking with Animals and Art.* Minneapolis: University of Minnesota Press, 2011.

Bryld, Mette, and Nina Lykke. *Cosmodolphins: Feminist Cultural Studies of Technology, Animals, and the Sacred.* London: Zed, 1999.

Burnett, D. Graham. *The Sounding of the Whale: Science and Cetaceans in the Twentieth Century.* Chicago: University of Chicago Press, 2012.

Butler, Octavia. *Kindred.* Garden City: Doubleday, 1979.

Butler, Octavia. *Lilith's Brood.* New York: Grand Central, 2000.

Cahill, James Leo. *Zoological Surrealism: The Nonhuman Cinema of Jean Painlevé.* Minneapolis: University of Minnesota Press, 2019.

Carson, Rachel L. *Under the Sea Wind.* New York: Simon & Schuster, 1941.

Carson, Rachel. *The Sea Around Us.* Oxford: Oxford University Press, 1951.

Castro, Peter, and Michael E. Huber. *Marine Biology.* 7th ed. Boston: McGraw-Hill, 2008.

Caws, Mary Ann, ed. *Surrealism.* London: Phaidon, 2004.

Chakrabarty, Dipesh. "The Climate of History." *Critical Inquiry* 35 (2009): 197–222.

Chang, David A. *The World and All the Things Upon It: Native Hawaiian Geographies of Exploration.* Minneapolis: University of Minnesota Press, 2016.

Clarke, Arthur C. *The Deep Range.* 1957. In *"The Ghost from the Grand Banks" and "The Deep Range."* New York: Warner Books, 2001.

Cohen, Margaret. "Denotation in Alien Environments: The Underwater Je Ne Sais Quoi." *Representations* 125, no. 1 (2014): 104–26.

Cohen, Margaret. *The Underwater Eye: How the Movie Camera Opened the Depths and Unleashed New Realms of Fantasy.* Princeton, N.J.: Princeton University Press, 2022.

Conniff, Richard. *The Species Seekers: Heroes, Fools, and the Mad Pursuit of Life on Earth.* New York: Norton, 2011.

Connors, A. *The Girl Who Broke the Sea*. London: Scholastic, 2023.

Cooley, Alison. "Listening for Sedna: Contemporary Inuit Art and Climate." *Inuit Art Quarterly,* September 30, 2016. https://www.inuitartfoundation.org/iaq-online/29-1-listening-for-sedna.

Cracking the Ocean Code. Directed by David Conover. Film. Discovery Communications, 2005.

Cressey, Daniel. "Marine Biology: Out of the Blue." *Nature* 467 (2010): 514–15.

Crist, Darlene Trew, Gail Scowcroft, and James M. Harding, Jr. *World Ocean Census.* Buffalo, N.Y.: Firefly, 2009.

Crumpton, Michael. Comments to YouTube video, "The Bathysphere: Discovering the Deep," July 17, 2020. https://www.youtube.com/watch?v=likHIvS5F18.

Csicsery-Ronay, Jr., Istvan. *The Seven Beauties of Science Fiction.* Middletown, Conn.: Wesleyan University Press, 2008.

Daley, Christopher. "The Not So Cozy Catastrophe: Reimaging the British Disaster Novel." In *Apocalyptic Discourse in Contemporary Culture: Post-millennial Perspectives on the End of the World,* edited by Monica Germana and Aris Mousoutzanis, 133–47. New York: Routledge, 2014.

Darby, Hugh. "A World Under Water." Review of William Beebe, *Half Mile Down. Nation,* December 12, 1934, 686–87.

Darwin, Charles. *The Origin of Species.* New York: P. F. Collier and Son, 1909.

Daston, Loraine J., and Peter Galison. *Objectivity.* New York: Zone, 2007.

"The Deep." *Blue Planet 2.* Directed by James Honeyborne. HBO Max, 2018.

The Deep. Exhibition. Bloom Association. https://www.bloomassociation.org/en/our-actions/our-activities/awareness-education/exhibition-into-the-deep/. Accessed March 24, 2024.

Deep Abyss. Game. Studio RO, 2022. Steam Store. https://store.steampowered.com/app/1299880/Deep_Abyss/.

"Deep Sea Creatures." Video. Census of Marine Life website. Accessed 2012.

Deep Sea (Shen hai). Written and directed by Xiaopeng Tian. Film. October Media, 2023.

De Loughrey, Elizabeth M. *Allegories of the Anthropocene.* Durham, N.C.: Duke University Press, 2016.

Derrida, Jacques. *The Animal That Therefore I Am.* Translated by David Wills. New York: Fordham, 2008.

Derrida, Jacques. *Margins of Philosophy.* Translated by Alan Bass. Chicago: University of Chicago Press, 1982.

Derrida, Jacques. *Of Grammatology.* Translated by Gayatri Chakravorty Spivak. Baltimore, Md.: Johns Hopkins University Press, 1976.

de Waal, Frans B. M. "Anthropomorphism and Anthropodenial: Consistency in Our Thinking about Humans and Other Animals." *Philosophical Topics* 27, no. 1 (1999): 255–80.

Dickinson, Emily. "The brain—is wider than the sky." Poem 598 in *The Poems of Emily Dickinson,* edited by R. W. Franklin. Cambridge, Mass.: Belknap, 2005.

Diolé, Philippe. *The Undersea Adventure.* Translated by Alan Ross. New York: Julian Messner, 1953.

Dion, Mark, Katherine McLeod, and Madeleine Thompson, curators. *Exploratory Works: Drawings from the Department of Tropical Research Field Expeditions.* Exhibition brochure. New York: Drawing Center, 2017.

Dive to the Edge of Creation. Directed by James Lipscomb and Alfred Giddings. Narrated by Leslie Nielsen. Film. National Geographic Special, 1979.

Doyle, Richard. *Darwin's Pharmacy: Sex, Plants, and the Evolution of the Noosphere.* Seattle: University of Washington Press, 2011.

Drayton, Richard. "Interview with Richard Drayton." Conducted by Katherine McLeod. In Dion, McLeod, and Thompson, *Exploratory Works,* 57–69.

Drexciya. *Deep Sea Dweller.* Album. Shockwave Records, 1992.

Earle, Sylvia. *Sea Change.* New York: Ballantine, 1996.

Ebbe, Brigitte, et al. "Diversity of Abyssal Marine Life." In *Life in the World's Oceans: Diversity, Distribution, and Abundance,* edited by Alasdair D. McIntyre, 139–60. Oxford: Wiley-Blackwell, 2010.

Elias, Ann. *Coral Empire: Underwater Oceans, Colonial Tropics, Visual Modernity.* Durham, N.C.: Duke University Press, 2019.

Enik, Ted. *The Bathysphere Boys: The Depth-Defying Diving of Beebe and Barton.* Atglen, Pa.: Schiffer Kids, 2019.

Escobar, Arturo. *Designs for the Pluriverse: Radical Interdependence, Autonomy, and the Making of Worlds.* Durham, N.C.: Duke University Press, 2017.

Fay, Jennifer. *Inhospitable World: Cinema in the Time of the Anthropocene.* Oxford: Oxford University Press, 2018.

Fessenden, Melissa. “In the Early 20th Century the Department of Tropical Research Was a Glamorous Adventure.” *Smithsonian,* April 25, 2017. https://www.smithsonianmag.com/arts-culture/early-20th-century-department-tropical-research-was-full-glamorous-adventure-180963020/.

Firebrace, William. *Memo for Nemo.* London: Architectural Association, 2016.

Fisher, Mark. *The Weird and the Eerie.* London: Repeater, 2016.

Flusser, Vilém, and Louis Bec. *Vampyroteuthis Infernalis.* Translated by Valentine A. Pakis. Minneapolis: University of Minnesota Press, 2012.

Fox, Brad. *The Bathysphere Book: Effects of the Luminous Ocean Depths.* New York: Astra House, 2023.

Gerspacher, Arnaud. *The Owls Are Not What They Seem: Artist as Ethologist.* Minneapolis: University of Minnesota Press, 2022.

Gilbert, Sandra, and Susan Gubar. *No Man’s Land: The Place of the Woman Writer in the Twentieth Century—The War of the Words.* New Haven, Conn.: Yale University Press, 1989.

Gómez-Barris, Macarena. *The Extractive Zone: Social Ecologies and Decolonial Perspectives.* Durham, N.C.: Duke University Press, 2017.

Gould, Alex, Alan Jamieson, and Thom Linley. “Aesthetics of the Deep Sea with Artist Alex Gould.” *Deep Sea Podcast,* September 3, 2020. https://www.armatusoceanic.com/podcast/003-aesthetics-of-thedeep-sea-with-artist-alex-gould.

Gould, Carol Grant. *The Remarkable Life of William Beebe, Explorer and Naturalist.* Washington, D.C.: Island Press, 2004.

Graham, Jorie. “Deep Water Trawling.” In *fast,* 6–7. New York: Harper Collins, 2017.

Grassle, Fred, and Nancy J. Maciolek. “Deep-Sea Species Richness: Regional and Local Diversity Estimates from Quantitative Bottom Samples.” *American Naturalist* 139, no. 2 (1992): 313–41.

Gumbs, Alexis Pauline. *Dub: Finding Ceremony.* Durham, N.C.: Duke University Press, 2020.

Gumbs, Alexis Pauline. *M Archive: After the End of the World.* Durham, N.C.: Duke University Press, 2018.

Gumbs, Alexis Pauline. *Undrowned: Black Feminist Lessons from Marine Mammals.* Chico, Calif.: A. K. Press, 2020.

“Hard to See.” Video coproduced with National Geographic. Census of Marine Life. http://www.coml.org/video-gallery. Accessed February 2, 2010.

“Hard to See Creatures.” Video coproduced with National Geographic. Census of Marine Life. http://www.coml.org/video-gallery. Accessed February 2, 2010.

Hamblin, Jacob. *Oceanographers and the Cold War: Disciplines of Marine Science*. Seattle: University of Washington Press, 2005.

Haraway, Donna. *The Companion Species Manifesto*. Chicago: Prickly Paradigm, 2003.

Haraway, Donna. *Primate Visions: Gender, Race, and Nature in the World of Modern Science*. New York: Routledge, 1989.

Haraway, Donna J. "Situated Knowledges: The Science Question in Feminism and the Privilege of Partial Perspective." In *Simians, Cyborgs, Women: The Reinvention of Nature*, 183–202. New York: Routledge, 1991.

Haraway, Donna J. *Staying with the Trouble: Making Kin in the Chthulucene*. Durham, N.C.: Duke University Press, 2016.

Hau'ofa, Epeli. "Our Sea of Islands." In *We Are the Ocean: Selected Works*, 27–40. Honolulu: University of Hawai'i Press, 2008.

Hayward, Eva. "Sensational Jellyfish: Aquarium Affects and the Matter of Immersion." *differences* 25, no. 3 (2012): 161–96.

Heise, Ursula. *Imagining Extinction: The Cultural Meanings of Endangered Species*. Chicago: University of Chicago, 2016.

Helmreich, Stefan. *Alien Ocean: Anthropological Voyages in Microbial Seas*. Berkeley: University of California Press, 2009.

Helmreich, Stefan, and Sophia Roosth. "Life Forms: A Keyword Entry." In *Sounding the Limits of Life: Essays in the Anthropology of Biology and Beyond*, edited by Stefan Helmreich. 19–34. Princeton, N.J.: Princeton University Press, 2016.

Horton, Zachary. *The Cosmic Zoom: Scale, Knowledge, Mediation*. Chicago: University of Chicago Press, 2021.

Idyll, Clarence P. *Abyss: The Deep Sea and the Creatures that Live in It*. New York: Thomas Y. Crowell, 1976.

Iheka, Cajetan. *African Ecomedia: Network Forms, Planetary Politics*. Durham, N.C.: Duke University Press, 2021.

Iman Jackson, Zakiyyah. *Becoming Human: Matter and Meaning in an Antiblack World*. New York: NYU Press, 2020.

"Indigenous Voices for a Ban on Deep Sea Mining." Petition. Blue Climate Initiative. https://www.blueclimateinitiative.org/say-no-to-deep-sea-mining.

Ingersoll, Karin Amimoto. *Waves of Knowing: A Seascape Epistemology*. Durham, N.C.: Duke University Press, 2016.

Ingram, Penelope. *Imperiled Whiteness: How Hollywood and Media Make Race in Postracial America*. Jackson: University of Mississippi Press, 2023.

Into the Deep. Exhibit. Monterey Bay Aquarium, Monterey, Calif., 2022.

Irigaray, Luce. *Marine Lover of Friedrich Nietzsche.* 1980. Translated by Gillian C. Gill. New York: Columbia University Press, 1991.

Jacobi, Christopher J., et al. "Aesthetic Experiences and Flourishing in Science: A Four-Country Study." *Frontiers in Psychology* 13 (2022): 923940.

Jamieson, Alan. *The Hadal Zone: Life in the Deepest Ocean.* Cambridge: Cambridge University Press, 2015.

Jamieson, Alan. "Meet the Mysterious 'Monsters' of the Deep Sea." TED Talk, October 31, 2022. https://www.ted.com/talks/alan_jamieson_meet_the_mysterious_monsters_of_the_deep_sea?language=en.

Jamieson, Alan J., Thomas D. Linley, and Prema Arasu. "Reply to: People Do Care about the Deep Sea: A Comment on Jamieson et al. (2020)." *ICES Journal of Marine Science* 79, no. 8 (2022): 2340–43.

Jamieson, Alan J., and Deo Florence L. Onda. "Lebensspuren and Müllspuren: Drifting Plastic Bags Alter Microtopography of Seafloor at Full Ocean Depth (10,000 m, Philippine Trench)." *Continental Shelf Research* 250 (2022): 1–6.

Jamieson, Alan J., Glenn Singleman, Thomas D. Linley, and Susan Casey. "Fear and Loathing of the Deep Ocean: Why Don't People Care about the Deep Sea." *ICES Journal of Marine Science* 78, no. 3 (2020): 797–809.

Jamieson, Alan J., et al. "Bioaccumulation of Persistent Organic Pollutants in the Deepest Ocean Fauna." *Natural Ecology and Evolution* 1, no. 3 (2017): 51.

Jeffords, Susan. *Hard Bodies: Hollywood Masculinity in the Reagan Era.* Newark, N.J.: Rutgers University Press, 1993.

Johnson, Elizabeth R., and Irus Braverman. "Blue Legalities: Governing More-than-Human Oceans." Introduction to *Blue Legalities: The Life and Laws of the Sea,* edited by Irus Braverman and Elizabeth R. Johnson, 1–24. Durham, N.C.: Duke University Press, 2020.

Jue, Melody. "Intimate Objectivity: On Nnedi Okorafor's Oceanic Afrofuturism." *Women's Studies Quarterly* 45 (2017): 171–88.

Jue, Melody. *Wild Blue Media: Thinking through Seawater.* Durham, N.C.: Duke University Press 2020.

Kaharl, Victoria A. *Water Baby: The Story of* Alvin. Oxford: Oxford University Press, 1990.

Ketterer, David. "John Wyndham: The Facts of Life Sextet." In *A Companion to Science Fiction,* edited by David Seed, 375–88. Oxford: Wiley-Blackwell, 2005.

Killers of the Sea. Film, 48 minutes, black and white. Directed by Ray Friedgen Productions, 1937.

Kimmerer, Robin Wall. "Speaking of Nature." *Orion Magazine,* March–April 2017. https://orionmagazine.org/article/speaking-of-nature/.

King, Tiffany Lethabo. *The Black Shoals: Offshore Formations of Black and Native Studies.* Durham, N.C.: Duke University Press, 2019.

Kingsland, Sharon E. *The Evolution of American Ecology,* 1890–2000. Baltimore: Johns Hopkins University Press, 2008.

Klein, Joanna. "They Mixed Science, Art, and Cocktail Parties to Reveal the Mysteries of the Sea." *New York Times,* March 27, 2017. https://www.nytimes.com/2017/03/27/science/william-beebe-department-of-tropical-research-illustrations.html.

Knowlton, Nancy. *Citizens of the Sea: Wondrous Creatures from the Census of Marine Life.* Washington, D.C.: National Geographic, 2010.

Kolodny, Annette. *The Lay of the Land: Metaphor as Experience and History in American Life and Letters.* Chapel Hill: University of NC Press, 1984.

Kornbluh, Anna. *Immediacy, or The Style of Too Late Capitalism.* London: Verso, 2023.

Koslow, Tony. *The Silent Deep: The Discovery, Ecology, and Conservation of the Deep Sea.* Sydney: UNSW Press, 2007.

Kraese, Danielle. *Deep-Sea Creeps: A Field Guide to Terrible Ex-Boyfriends (As Sea Creatures).* Melbourne: Smith Street, 2024.

Krause, Kea. "Scales Too Large, Dorsal Rays Too Few." Lenny, October 4, 2018. https://www.lennyletter.com/story/scales-too-large-dorsal-rays-too-few.

Kroll, Gary. *America's Ocean Wilderness: A Cultural History of Twentieth-Century Exploration.* Lawrence: University Press of Kansas, 2008.

Kroll, Gary. "William Beebe." In *Dictionary of Literary Biography Volume 275: Twentieth-Century American Nature Writers: Prose,* 39–47. Detroit: Gale, 2008.

LaFollette, Marcel C. *Making Science Our Own: Public Images of Science, 1910–1955.* Chicago: University of Chicago Press, 1990.

Latour, Bruno. "An Attempt at a 'Compositionist Manifesto.'" *New Literary History* 41, no. 3 (2010): 471–90.

Latour, Bruno. *On the Modern Cult of the Factish Gods.* Durham, N.C.: Duke University Press, 2010.

Latour, Bruno. *Pandora's Hope: Essays on the Reality of Science Studies.* Cambridge, Mass.: Harvard University Press, 1999.

Latour, Bruno. *We Have Never Been Modern.* Translated by Catherine Porter. Cambridge, Mass.: Harvard University Press, 1993.

Latour, Bruno. "Why Critique Has Run Out of Steam? From Matters of Fact to Matters of Concern." *Critical Inquiry* 30 (2004): 225–48.

Ledgard, J. M. *Submergence.* Minneapolis: Coffee House, 2013.

Leyssen, Jan. *Mission to the Bottom of the Sea.* Hasselt, Belgium: Clavis, 2020.

"*Lophodolos acanthognathus* Regan, 1925." World Register of Marine Species. https://www.marinespecies.org/aphia.php?p=taxdetails&id=126567.

Mann, Bonnie. *Women's Liberation and the Sublime: Feminism, Postmodernism, Environment.* Oxford: Oxford University Press, 2006.

Marc, Christopher. "Director Bong Joon Ho Says His Animated Film Won't Release until 2025–2026." Ronin, July 8, 2021. https://theronin.org/2021/07/08/director-bong-joon-ho-says-his-animated-film-wont-release-until-2025-2026-inspired-by-the-deep-the-extraordinary-creatures-of-the-abyss/.

Marder, Michael. *Plant Thinking: A Philosophy of Vegetal Life.* New York: Columbia University Press, 2013.

"Mariana Trench First-Ever Look: Cameron Releases Video from 7 Mile Deep Abyss." Video, RT News, March 27, 2012. https://www.rt.com/news/mariana-trench-james-cameron-first-video-608/.

Matsen, Brad. *Descent: The Heroic Discovery of the Abyss.* New York: Vintage, 2005.

Mawani, Renisa. *Across Oceans of Law: The Komagata Maru and Jurisdiction in the Time of Empire.* Durham, N.C.: Duke University Press, 2018.

McIntyre, Alasdair, ed. *Life in the World's Oceans: Diversity, Distribution, and Abundance.* Census of Marine Life. Oxford: Wiley-Blackwell, 2010.

McKittrick, Katherine. *Dear Science and Other Stories.* Durham, N.C.: Duke University Press, 2021.

McLeod, Katherine. "Sinking beneath the Surface—William Beebe, the Department of Tropical Research, and Marine Biology." *Bulletin, Association for the Sciences of Limnology and Oceanography* 24, no. 2 (2015): 26–31.

McVeigh, Karen. "More than 5,000 New Species Discovered in Pacific Deep Sea Mining Hotspot." *Guardian,* May 25, 2023. https://www.theguardian.com/environment/2023/may/25/more-than-5000-new-species-discovered-in-pacific-deep-sea-mining-hotspot.

Mead, Margaret. "Patterns of Culture." Review of Ruth Benedict, *Patterns of Culture. Nation,* December 12, 1934, 686.

Meg 2: The Trench. Directed by Ben Wheatley. CMC Pictures, 2023.

"Meet the Lumpsucker." Ocean Conservancy, December 27, 2019. https://oceanconservancy.org/blog/2019/12/27/lumpsucker/. Accessed March 2024.

Mentz, Steve. *At the Bottom of Shakespeare's Ocean.* New York: Continuum, 2009.

Merchant, Carolyn. *The Death of Nature: Women, Ecology, and the Scientific Revolution.* New York: Harper Collins, 1990.

Mitchell, W. J. T. *What Do Pictures Want? The Lives and Loves of Images.* Chicago: University of Chicago Press, 2005.

Mitman, Greg. *Reel Nature: America's Romance with Wildlife on Film.* Cambridge, Mass.: Harvard University Press, 1999.

Montemurno, Megan. "Meet the Giant Isopod." Ocean Conservancy, September 21, 2023. https://oceanconservancy.org/blog/2023/09/21/meet-giant-isopod-deep-sea/.

Moorhead, Joanna. "Dalí in a Diving Helmet: How the Spaniard Almost Suffocated Bringing Surrealism to England." *Guardian,* June 1, 2016. https://www.theguardian.com/artanddesign/2016/jun/01/dali-exhibition-surreal-encounters-edinburgh.

Moreton-Robinson, Aileen. *The White Possessive: Property, Power, and Indigenous Sovereignty.* Minneapolis: University of Minnesota Press, 2015.

Morgan, Nigel, composer. "Sounding the Deep." 2012. Words by Phil Legard and Nigel Morgan.

Mukerji, Chandra. *A Fragile Power: Scientists and the State.* Princeton, N.J.: Princeton University Press, 1989.

My Octopus Teacher. Directed by James Reed and Pippa Erlich. Film. Netflix, 2020.

Nabrit, Samuel. "Samuel Nabrit, 98, Scientist and a Pioneer in Education." Obituary by N. Batric Kimetris. *New York Times,* January 6, 2004.

Nadeau, Maurice. *The History of Surrealism.* Translated by Richard Howe. New York: Macmillan, 1965.

Nagel, Thomas. "What Is It Like to Be a Bat?" *Philosophical Review* 83, no. 4 (1974): 435–50.

Neimanis, Astrida. "Held in Suspense: Mustard Gas Legalities in the Gotland Deep." In *Blue Legalities: The Life and Laws of the Sea,* edited by Irus Braverman and Elizabeth R. Johnson, 45–62. Durham, N.C.: Duke University Press, 2020.

Nixon, Rob. *Slow Violence and the Environmentalism of the Poor.* Cambridge, Mass.: Harvard University Press, 2011.

Norris, Margot. *Beasts of the Modern Imagination.* Baltimore, Md.: John Hopkins University Press, 2019.

Nouvian, Claire. *The Deep: The Extraordinary Creatures of the Abyss.* Chicago: University of Chicago Press, 2007.

Nouvian, Claire. "Meet Claire Nouvian." Goldman Environmental Prize, 2018. https://www.goldmanprize.org/recipient/claire-nouvian/#recipient-bio.

Okorafor, Nnedi. *Lagoon.* New York: Saga, 2014.

Oreskes, Naomi. "Scaling Up Our Vision." *Isis* 105, no. 2 (2014): 379–91.

Oreskes, Naomi. *Science on a Mission: How Military Funding Shaped What We Do and Don't Know about the Ocean.* Chicago: University of Chicago Press, 2021.

Osborn, Henry Fairfield. "A New Method of Deep Sea Observation at First Hand." *Science,* July 11, 1930, 28.

Pickering, Andrew. *The Mangle of Practice: Time, Agency, and Science.* Chicago: University of Chicago, 1995.

Pollan, Michael. *How to Change Your Mind.* Netflix TV series, 2022.

Poltisch, Kent E. *Beebe and Bostelmann.* Self-published, 2021.

Pound, Ezra. "A Retrospect." In *Literary Essays of Ezra Pound,* 12. New York: New Directions, 1986.

Puig de la Bellacasa, María. *Matters of Care: Speculative Ethics in More Than Human Worlds.* Minneapolis: University of Minnesota Press, 2017.

Pyle, Forest. *Art's Undoing: In the Wake of a Radical Aestheticism.* New York: Fordham University Press, 2014.

Quashie, Kevin. *Black Aliveness, or A Poetics of Being,* Durham, N.C.: Duke University Press, 2021.

Radomska, Marietta, and Cecilia Åsberg. "Fathoming Postnatural Oceans: Towards a Low Trophic Theory in the Practices of Feminist Posthumanities." *Environment and Planning E: Nature and Space* 5, no. 3 (2022): 1428–45.

Ray Troll and the Ratfish Wranglers. "Ballad of William Beebe." *Where the Fins Meet the Frets,* audio, 2008.

Reid, Susan. "Solwara 1 and the Sessile Ones." In *Blue Legalities: The Life and Laws of the Sea,* edited by Irus Braverman and Elizabeth R. Johnson, 25–44. Durham, N.C.: Duke University Press, 2020.

Rex, Michael A., and Ron J. Etter. *Deep-Sea Biodiversity: Pattern and Scale.* Cambridge, Mass.: Harvard University Press, 2010.

Roberts, Donna. "Surrealism and Natural History: Nature and the Marvelous in Breton and Caillois." In *A Companion to Dada and Surrealism,* edited by David Hopkins, 287–303. Oxford: Wiley-Blackwell, 2016.

Roorda, Eric Paul, ed. *The Ocean Reader: History, Culture, Politics.* Durham, N.C.: Duke University Press, 2020.

Rosenstock, Barbara. *Otis and Will Discover the Deep.* Boston: Little, Brown, 2018.

Rozwadowski, Helen M. *Fathoming the Ocean: The Discovery and Exploration of the Deep Sea.* Cambridge, Mass.: Harvard University Press, 2005.

San Diego Supercomputer Center. "Roger Arlinger Young." San Diego Supercomputer Center Presents Women in Science. https://www.sdsc.edu/ScienceWomen/index.html.

Saxon, Charles. Untitled cartoon captioned "I don't know why I don't care about the bottom of the ocean, but I don't." *New Yorker,* March 21, 1983.

Scales, Helen. *The Brilliant Abyss: Exploring the Majestic Hidden Life of the*

Deep Ocean and the Looming Threat That Imperils It. New York: Atlantic Monthly Press, 2021.

Schaffner, Ingrid. *Salvador Dalí's Dream of Venus: The Surrealist Funhouse from the 1939 World's Fair.* New York: Princeton Architectural Press, 2002.

Schlee, Susan. *A History of Oceanography: The Edge of An Unfamiliar World.* London: Robert Hale, 1975.

Seager, Joni. "Radical Observation." *Women's Studies Quarterly* 45, no. 1–2 (2017): 269–77.

Sehgal, Melanie. "Aesthetic Concerns, Philosophical Fabulations: The Importance of a 'New Aesthetic Paradigm.'" *Substance* 47, no. 1 (2018): 112–29.

Serres, Michel. *Malfeasance: Appropriation through Pollution.* Translated by Anne-Marie Feenberg-Dibon. Stanford, Calif.: Stanford University Press, 2011.

Shapin, Steven, and Simon Schaffer. *Leviathan and the Air-Pump.* Princeton, N.J.: Princeton University Press, 1985.

Shaviro, Steven. *Without Criteria: Kant, Whitehead, Deleuze, and Aesthetics.* Cambridge, Mass.: MIT Press, 2009.

Sheldon, David. *Into the Deep: The Life of Naturalist and Explorer William Beebe.* Watertown, Mass.: Charlesbridge, 2009.

Shiva, Vandana. *Biopiracy: The Plunder of Nature and Knowledge.* Boston: South End, 1997.

Shreeve, James. "Craig Venter's Epic Voyage to Redefine the Origin of the Species." *Wired,* August 1, 2004. http://www.wired.com/wired/archive/12.08/venter.html.

Shukin, Nicole. *Animal Capital: Rendering Life in Biopolitical Times.* Minneapolis: University of Minnesota Press, 2009.

Silverman, Kaja. *Miracle of Analogy, or The History of Photography, Part I.* Stanford, Calif.: Stanford University Press, 2015.

Silverman, Kaja. *World Spectators.* Stanford, Calif.: Stanford University Press, 2000.

Snelgrove, Paul V. R. *Discoveries of the Census of Marine Life: Making Ocean Life Count.* Cambridge: Cambridge University Press, 2010.

Snelgrove, Paul V. R., and Craig Randall Smith. "A Riot of Species in an Environmental Calm: The Paradox of the Species-Rich Deep Sea Floor." *Marine Biology Annual Review* 40 (2002): 311–42.

Snitow, Ann. "A Gender Diary." In *Conflicts in Feminism,* edited by Marianne Hirsch and Evelyn Fox Keller, 9–43. New York: Routledge, 1990.

Starosielski, Nicole. "Depth Mediators: Undersea Cables, Network Infrastructure and the Deep Ocean." In *Deep Mediations: Thinking Space in Cinema and Digital Cultures,* edited by Karen Redrobe and Jeff Scheible, 262–85. Minneapolis: University of Minnesota Press, 2021.

Steinberg, Philip E. *The Social Construction of the Ocean.* Cambridge: Cambridge University Press, 2001.

Stengers, Isabelle. "The Challenge of Ontological Politics." In *A World of Many Worlds,* edited by Marisol de la Cadena and Mario Blaser, 83–111. Durham, N.C.: Duke University Press 2018.

Stengers, Isabelle. *Thinking with Whitehead: A Free and Wild Creation of Concepts.* Translated by Michael Chase. Cambridge, Mass.: Harvard University Press, 2011.

Styrsky, Jindrich. "The Cuttlefish Man." 1934. In *History of the Surrealist Movement,* edited by Gérard Durozoi and translated by Alison Anderson. Chicago: University of Chicago Press, 2002.

Submergence. Directed by Wim Wenders. Film. Samuel Goldwyn Films, 2017.

Tee-Van, Helen Damrosch. *"Lophodolus."* 1929. WCS Archives, WCS-1039-01-05-0060-R-CR.

Teevee, Ningiukulu (Kinngait). "Sedna's Creation." 2019. https://steinbruecknativegallery.com/sednas-creation-ningiukulu-teevee/.

Teevee, Ningiukulu (Kinngait). "Sedna's Wonder." Cape Dorset, 2009. https://inuit.com/products/pp170309.

Teevee, Ningiukulu (Kinngait). "Untitled (Sedna by the Sea)." Ink on paper, 2001–2. In Alison Cooley, "Listening for Sedna: Contemporary Inuit Art and Climate," *Inuit Art Quarterly,* September 30, 2016. https://www.inuitartfoundation.org/iaq-online/29-1-listening-for-sedna.

Tennyson, Alfred. "The Kraken." 1830, https://poets.org/poem/kraken. Reprinted in John Wyndham, *The Kraken Wakes.* London: Penguin, 1953.

Thomas, Calvin. "Moments of Productive Bafflement, or Defamiliarizing Graduate Studies in English." *Pedagogy* 5, no. 1 (2005): 19–35.

Tuck, E., and K. W. Yang. "Decolonization Is Not a Metaphor." *Decolonization* 1, no. 1 (2012): 1–40.

"Ultraviolet Rays Make Rare Fish Transparent." *Science News-Letter* 26, no. 714 (December 15, 1934): 375.

Van Dover, Cindy Lee. "Deep in the Sea, Imagining the Cradle of Life on Earth: An Interview with Cindy Lee Van Dover." Conducted by Claudia Dreyfus. *New York Times,* October 16, 2007. https://www.nytimes.com/2007/10/16/science/16conv.html.

Van Dover, Cindy Lee. "Hydrothermal Vent Ecosystems and Conservation." *Oceanography* 25, no. 1 (2012): 313–16.

Van Dover, Cindy Lee. *Deep-Ocean Journeys: Discovering New Life at the Bottom of the Sea.* New York: Perseus, 1996.

Vecchione, Michael, Louise Allcock, Imants Priede, and Hans van Haren. *The Deep Ocean: Life in the Abyss.* Princeton, N.J.: Princeton University Press, 2023.

Vermeulen, Niki. "From Darwin to the Census of Marine Life: Marine Biology as Big Science." *PLoS One* 8, no. 1 (2013): e54284.

Verne, Jules. *Twenty Thousand Leagues under the Sea.* 1869–70. New York: Simon & Schuster, 2005.

Vint, Sherryl. *Animal Alterity: Science Fiction and the Question of the Animal.* Liverpool: Liverpool University Press, 2010.

Vitale, Francesco. *Biodeconstruction: Jacques Derrida and the Life Sciences.* Translated by Mauro Senatore. Albany: SUNY University Press, 2018.

von Uexküll, Jakob. *A Foray into the Worlds of Animals and Humans.* Translated by Joseph D. O'Neil. Minneapolis: University of Minnesota, 2010.

Voon, Claire. "The Colonial Legacy of a Globe Trotting Team of Artists and Scientists." *Hyperallergic,* July 11, 2017. https://hyperallergic.com/382055/exploratory-works-drawings-from-the-department-of-tropical-research-field-expeditions-drawing-center/.

Wall, Clare. "Here Be Monsters: Posthuman Adaptation and Subjectivity in Peter Watts' *Starfish.*" In *The Canadian Fantastic in Focus: New Perspectives,* edited by Allan Weiss, 67–80. Jefferson, N.C.: McFarland, 2015.

Walls, Laura Dassow. *The Passage to Cosmos: Alexander Von Humbolt and the Shaping of America.* Chicago: University of Chicago Press, 2009.

Warren, Calvin L. *Ontological Terror: Blackness, Nihilism, and Emancipation.* Durham, N.C.: Duke University Press, 2018.

Watkins, Claire Vaye. *Gold Fame Citrus: A Novel.* New York: Riverhead, 2016.

Watts, Peter. *Starfish.* New York: Doherty, 1999.

Watts, Peter. "Wildlife: Natural and Artificial. An Interview with Peter Watts." Conducted by Imre Szeman and Maria Whiteman. *Extrapolation* 48, no. 3 (2007): 603–19.

(WCS Archives) Wildlife Conservation Society, Department of Tropical Research Archive, Bronx Zoo, New York. https://wcsarchives.libraryhost.com/.

Weaver, Jace. *The Red Atlantic: American Indigenes and the Making of the Modern World, 1000–1927.* Chapel Hill: University of North Carolina Press, 2014.

Weinberg, Alvin M. *Reflections on Big Science.* Cambridge, Mass.: MIT Press, 1967.

Weston, Johanna J. N., Priscilla Carrillo-Barragan, Thomas D. Linley, William D. K. Reid, and Alan J. Jamieson. "New Species of *Eurythenes* from Hadal Depths of the Mariana Trench, Pacific Ocean (Crustacea: Amphipoda)." *Zootaxa* 4748 (2020): zootaxa.4748.1.9.

Whitehead, Alfred North. *Adventures of Ideas.* New York: Free Press, 1933.

Whitehead, Alfred North. *Science and the Modern World.* New York: Free Press, 1925.

Widder, Edith. *Below the Edge of Darkness: A Memoir of Exploring Light and Life in the Deep Sea.* New York: Random House, 2021.

Widder, Edith. "The Fine Art of Exploration." *Oceanography* 29, no. 4 (2016): 170–77.

Wilkes, Tabea. "'Poison Rainbows': The Speculative Jurisdiction of Okorafor's *Lagoon.*" *Comparative Critical Studies* 19, no. 3 (2022): 299–313.

Wolfe, Cary. *Before the Law: Humans and Other Animals in a Biopolitical Frame.* Chicago: University Chicago Press, 2013.

Wolfe, Cary. *Ecological Poetics, or Wallace Stevens's Birds.* Chicago: University of Chicago Press, 2020.

Wolfe, Cary. "Jagged Ontologies in the Anthopocene, or The 5 Cs." In *Life in the Posthuman Condition: Critical Responses to the Anthropocene,* edited by S. E. Wilmer and Audronė Žukauskaitė, 195–221. Edinburgh: Edinburgh University Press, 2023.

Wolfe, Cary. *What Is Posthumanism?* Minneapolis: University of Minnesota Press, 2010.

Womack, Ytasha L. *Afrofuturism: The World of Black Sci-Fi and Fantasy Culture.* Chicago: Lawrence Hill, 2013.

Wyndham, John. *The Kraken Wakes.* London: Penguin, 1953.

Youatt, Rafi. *Counting Species: Biodiversity in Global Environmental Politics.* Minneapolis: University of Minnesota Press, 2010.

Young, Josh. *Expedition Deep Ocean.* New York: Pegasus, 2020.

Index

Page numbers in italic refer to figures.

(continued from page ii)

59 *The Probiotic Planet: Using Life to Manage Life*
Jamie Lorimer

58 *Individuation in Light of Notions of Form and Information, Volume* II, *Supplemental Texts*
Gilbert Simondon

57 *Individuation in Light of Notions of Form and Information*
Gilbert Simondon

56 *Thinking Plant Animal Human: Encounters with Communities of Difference*
David Wood

55 *The Elements of Foucault*
Gregg Lambert

54 *Postcinematic Vision: The Coevolution of Moving-Image Media and the Spectator*
Roger F. Cook

53 *Bleak Joys: Aesthetics of Ecology and Impossibility*
Matthew Fuller and Olga Goriunova

52 *Variations on Media Thinking*
Siegfried Zielinski

51 *Aesthesis and Perceptronium: On the Entanglement of Sensation, Cognition, and Matter*
Alexander Wilson

50 *Anthropocene Poetics: Deep Time, Sacrifice Zones, and Extinction*
David Farrier

49 *Metaphysical Experiments: Physics and the Invention of the Universe*
Bjørn Ekeberg

48 *Dialogues on the Human Ape*
Laurent Dubreuil and Sue Savage-Rumbaugh

47 *Elements of a Philosophy of Technology: On the Evolutionary History of Culture*
Ernst Kapp

46 *Biology in the Grid: Graphic Design and the Envisioning of Life*
Phillip Thurtle

45 *Neurotechnology and the End of Finitude*
Michael Haworth

44 *Life: A Modern Invention*
Davide Tarizzo

43 *Bioaesthetics: Making Sense of Life in Science and the Arts*
Carsten Strathausen

42 *Creaturely Love: How Desire Makes Us More and Less Than Human*
Dominic Pettman

41 *Matters of Care: Speculative Ethics in More Than Human Worlds*
María Puig de la Bellacasa

40 *Of Sheep, Oranges, and Yeast: A Multispecies Impression*
Julian Yates

39 *Fuel: A Speculative Dictionary*
Karen Pinkus

38 *What Would Animals Say If We Asked the Right Questions?*
Vinciane Despret

37 *Manifestly Haraway*
Donna J. Haraway

36 *Neofinalism*
Raymond Ruyer

35 *Inanimation: Theories of Inorganic Life*
David Wills

34 *All Thoughts Are Equal: Laruelle and Nonhuman Philosophy*
John Ó Maoilearca

33 *Necromedia*
Marcel O'Gorman

32 *The Intellective Space: Thinking beyond Cognition*
Laurent Dubreuil

31 *Laruelle: Against the Digital*
Alexander R. Galloway

30 *The Universe of Things: On Speculative Realism*
Steven Shaviro

29 *Neocybernetics and Narrative*
Bruce Clarke

28 *Cinders*
Jacques Derrida

27 *Hyperobjects: Philosophy and Ecology after the End of the World*
Timothy Morton

26 *Humanesis: Sound and Technological Posthumanism*
David Cecchetto

25 *Artist Animal*
Steve Baker

24 *Without Offending Humans: A Critique of Animal Rights*
Élisabeth de Fontenay

23 *Vampyroteuthis Infernalis: A Treatise, with a Report by the Institut Scientifique de Recherche Paranaturaliste*
Vilém Flusser and Louis Bec

22 *Body Drift: Butler, Hayles, Haraway*
Arthur Kroker

21 *HumAnimal: Race, Law, Language*
Kalpana Rahita Seshadri

20 *Alien Phenomenology, or What It's Like to Be a Thing*
Ian Bogost

19 *CIFERAE: A Bestiary in Five Fingers*
Tom Tyler

18 *Improper Life: Technology and Biopolitics from Heidegger to Agamben*
Timothy C. Campbell

17 *Surface Encounters: Thinking with Animals and Art*
Ron Broglio

16 *Against Ecological Sovereignty: Ethics, Biopolitics, and Saving the Natural World*
Mick Smith

15 *Animal Stories: Narrating across Species Lines*
Susan McHugh

14 *Human Error: Species-Being and Media Machines*
Dominic Pettman

13 *Junkware*
Thierry Bardini

12 *A Foray into the Worlds of Animals and Humans,* with *A Theory of Meaning*
Jakob von Uexküll

11 *Insect Media: An Archaeology of Animals and Technology*
Jussi Parikka

10 *Cosmopolitics II*
Isabelle Stengers

9 *Cosmopolitics I*
Isabelle Stengers

8 *What Is Posthumanism?*
Cary Wolfe

7 *Political Affect: Connecting the Social and the Somatic*
John Protevi

6 *Animal Capital: Rendering Life in Biopolitical Times*
Nicole Shukin

5 *Dorsality: Thinking Back through Technology and Politics*
David Wills

4 *Bíos: Biopolitics and Philosophy*
Roberto Esposito

3 *When Species Meet*
Donna J. Haraway

2 *The Poetics of DNA*
Judith Roof

1 *The Parasite*
Michel Serres

Stacy Alaimo is Barbara and Carlisle Moore Professor in English and core faculty member in environmental studies at the University of Oregon. She is author of *Undomesticated Ground: Recasting Nature as Feminist Space; Bodily Natures: Science, Environment, and the Material Self*; and *Exposed: Environmental Politics and Pleasures in Posthuman Times* (Minnesota, 2016).